Further Praise for

Little Deaths All in a Row

"Shades of Carmen Machado's *In the Dream House*, and I mean that as a galactic compliment. What an intriguing, compelling helix. Intellectually exciting and driven by a kind of heart-quest. SO THRILLING to see the 'mono story' of motherhood and marriage and sex blown to bits from the get-go."

— Lidia Yuknavitch, author of *The Chronology of Water*

"This vulnerable, cerebral, emotional, and powerful work will take you to unexpected, magical, and utterly mysterious places."

— Emily Rapp Black, author of *The Still Point of the Turning World*

"Juxtaposing paradoxes to collapse false binaries between birth and death, desire and grief, and mystery and knowing, Elizabeth Earley's *Little Deaths All In A Row* mines the veins opened by Maggie Nelson's *The Argonauts*. Earley deploys her own damage as a portal — a collective reckoning — to what is made possible when we choose to remain soft through hard looking. Plumbing the vectors of time, intimacy, and longing, these essays are as nonlinear, recurrent, and vital as healing. You can be made better by reading this book."

— Kristen Millares Young, award winning journalist and essayist and author of the novel, *Subduction*

LITTLE DEATHS ALL IN A ROW

ESSAYS ON SEX AND DEATH

BY ELIZABETH EARLEY

ILLUSTRATED BY NICOLE ROBERTS

ISBN: 9781938841439

Cover design and interior illustrations by Nicole Roberts

This book is available in paperback and electronic book format.

Earley, Elizabeth

Little Deaths All in a Row

For you, my love.

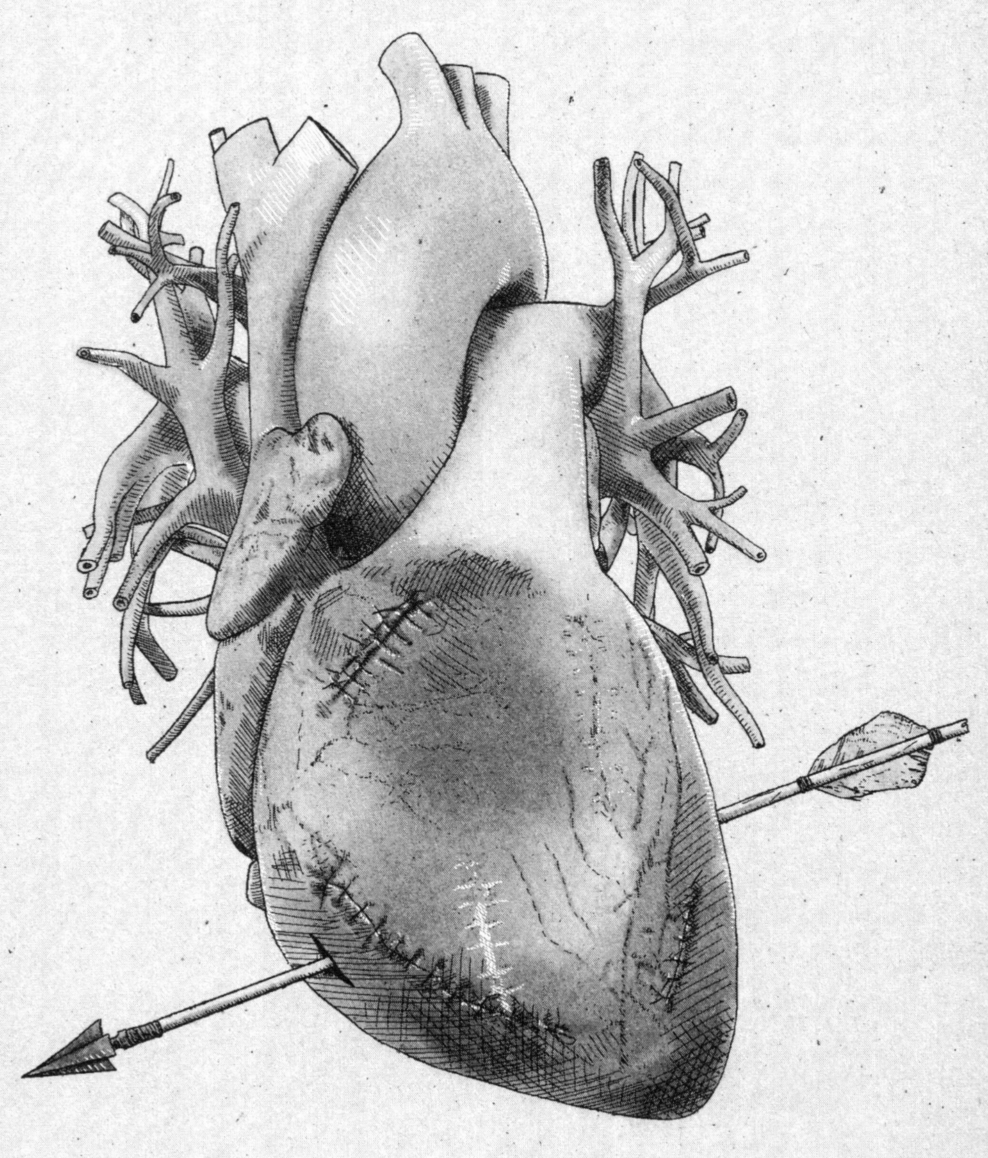

Invitation

Take me by the hands and come with me. Break the spine of this book with me. We will travel through time. Eons of time, elastic time, curved time. We will travel through space and anti-space, through vistas of thought you will recognize as familiar, as you have inhabited them already, before now. You have thought about love, you have thought about sex, you have thought about death and bodies and matter and what's real. You have likely thought about physics and biology and mythology and philosophy. All of it. Everything. I am taking you there, through everything and back again, right up to the cliff edge of life where you can feel your breath against that thinnest of veils. I am taking you through that veil to the other side where we will float, leaf-like, above all that's beyond. Where we can glimpse what's next. Together. Trust me.

To understand where I'm taking you, metaphors are necessary. Think of houses. Think large, spacious, many-roomed houses with hidden corridors and trap doors and staircases leading to nowhere. Hold them in your mind. Houses. Homes.

Fair warning: these houses are haunted. They contain malevolent energies, but also benevolent ones. It will help you to bear in mind as we go that underneath ideas of bad and good, there is only energy — strong and undifferentiated energy.

This journey is not for the faint of heart. It's definitely not for prudes.

Come if you're brave. Come if you're scared. Come if you're shy.

Come if you're at least a little slutty.

Hell, come if you are a prude. We're all going to die anyway, why not take a risk?

Come if you've trespassed and been forgiven.

Come if you can't always forgive yourself.

Come if you've trusted and been fucked over.

Come if the silver thread holding together the dazzling shards of your impossibly resilient broken and broken and broken again heart is grief. And if the gold thread is joy. And the blue thread is blissfully blind hope. And if the whole thing beats with fear.

Come if you have noticed the similarities between pleasure and pain, between penises and vaginas, between male and female, between feminine and masculine, between hetero and homo, between birth and death.

Come if you have held these binaries in the bowl of your hands, between the sacred and the profane, like distinct species of hinged, shelled creatures. Creatures carried in by the mischief of the moon and its influence on water, the dependable patterns making a round rhythm of the tides. Creatures like scallops, whelks, oysters, clams, mollusks — spiney and toothy bivalves with soft, wet insides. And holding them, these binaries, you studied them, memorized their names, lived by them — only to have them crumble and fall like dust through your fingers.

Come if you love bodies of water and if you've realized that you, yourself, are a body of water, and if part of what you love about these mysterious depths holding glittering creatures is the paradoxes they are — how they're comforting and healing and violent and deadly all at once.

Come if you're afraid you'll drown.

Importantly, come if you've been hurt and then healed badly, emerging disfigured with scars, and if you can sometimes see how these have made you more beautiful. How you have been made better by the pain. Improved by the ugly.

I'll be telling you some epic stories. We'll act them out together. If you'll indulge me in this role-play, you will be all the meaningful lovers I've had and lost. I invite you to give it your all. Even if there are cherished things you stand to lose, I'm asking you to risk it.

And if you take it with me, this journey, you will never be the same. Nor will I. And we will be connected. Always. Inextricably.

THE HOUSE OF LOVE

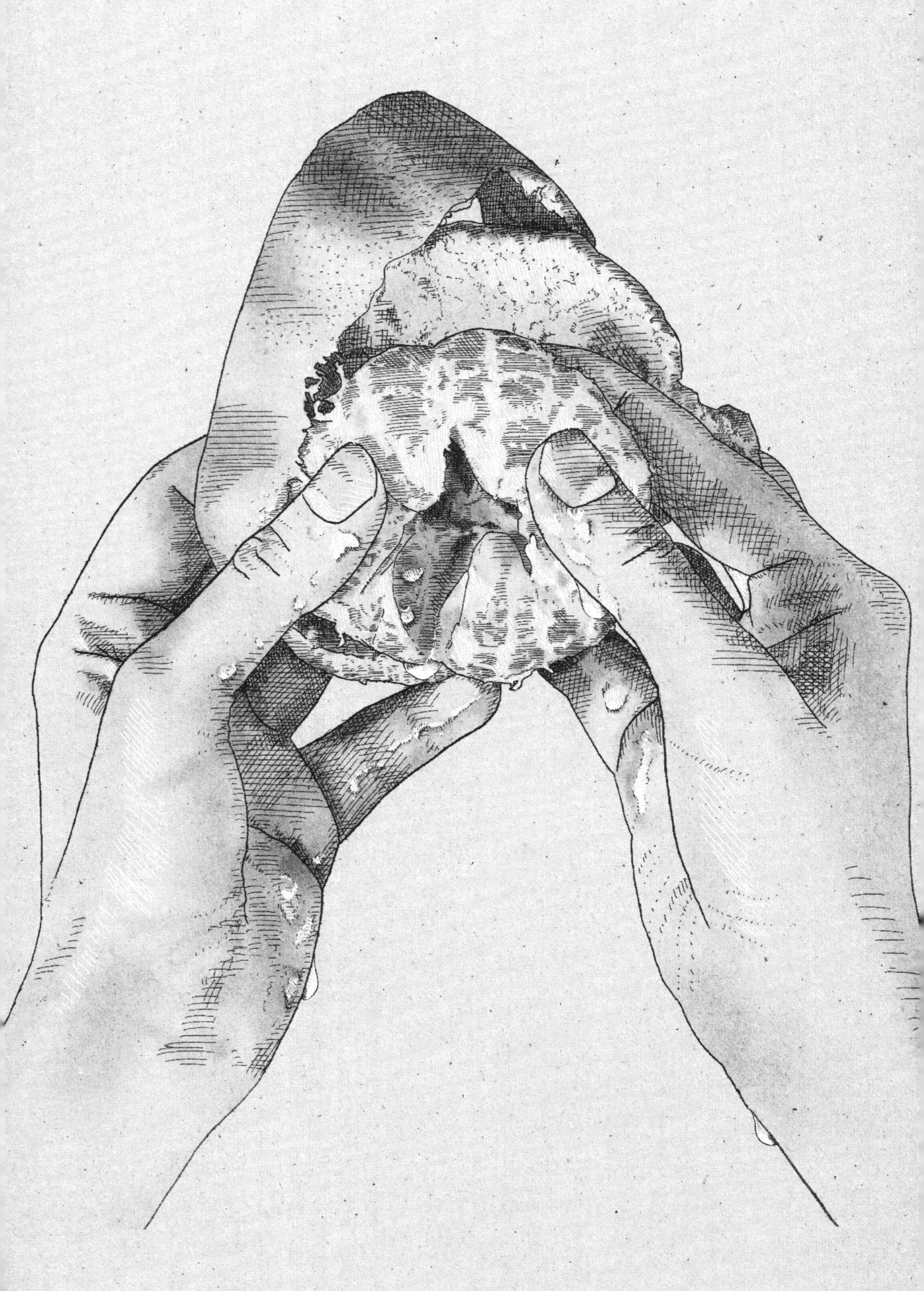

Undying

"Love is just love. It either has kissing and orgasms, or no kissing and orgasms."

You say these words and with the sound of your voice into my ears comes the smell of you into my nose. The smell there on your skin, stronger the closer I come, which I tell you jokingly is like the hospital. There is something about the smell of your skin that does remind me of the smell of the hospital I grew up in, where my sister lay comatose for a year then emerged a new version of herself created by the brain damage. But perhaps how you smell reminds me of that because it smells like home. Like everywhere I have ever wanted to be and everything I have ever wanted to eat and all the surfaces I have been drawn to touch with my hands.

"No. Love is not *just* love. There are dimensions. Like, hot dogs are great, right? Everyone loves a good hot dog. But have you ever had cheese on your hot dog? I mean…" I lift my hands and bring them near your face, as if to hold it, but then I close them into fists and give a small double punch to the air for emphasis. "It's a whole other physics. It's the *cheese* on the hot dog that compels the kissing and orgasms."

You smile and ask, "What have you been reading now?" And I feel it, that empty-yet-full feeling all the way down and all the way up my body.

You are my first love.

I've always remembered us having this conversation, but I can never place us in time and space when I remember it. I see us sitting by the ocean, but we were never anywhere near an ocean during our time together.

You are seventeen, and seventeen months older. One year ahead of me in school, I first notice you when I am a sophomore and you are a junior. We have choir together. From my vantage point at the top of the risers in the soprano section, I look straight at you standing at the top of the risers in the alto section. The curly, black hair floating around your face. The dark eyes and gold-toned skin. That cut of your profile when you turn your face away, mouth open in song, the curves and the planes and the shadow and the light gathering around the glistening insides of your open mouth — I am mesmerized. This is swoon before I know to call it that. My heart picks up pace, my body slightly trembles, and butterfly wings beat the air and the membranous walls of my abdominal cavity whenever we gain proximity. In my mind, we sing only to and for each other. Months pass before I gain the courage to talk to you, and when I do, the words sting my cheeks with hot shame the moment they tumble from my mouth to your perfect ears.

///

June was eighty-nine years old when I met her, inside six weeks of her death. Her ninetieth birthday was three months away, and she alternated between wanting to greet it and being ready to bow out sooner rather than later. Among the first dozen words she spoke to me were *I love you.* Not only was there a lack of hesitation in reciprocating her words, I emphasized the phrase by taking both her hands in mine and leaning toward her bright, wide-open face. Everything in her — her whole life's experiences, all the love and all the pain — it was there in that face, skin like fragile crumpled tissue paper with sky-blue eyes. There was something behind her eyes and under her skin, some unmistakable aliveness I recognized as ancient and wise, like an ocean current pulling the deep and pushing the shallow, sending it leaping from itself and crashing back into itself. And like her face, her heart was as open as they come. Just like a toddler, or a puppy on its back offering up its soft underbelly. It was in the openness of her face and the way she held my hand, kept saying she loved me.

Her husband, Clay, shuffled quietly across the room from his resident armchair, bent sharply forward as he went, nose pointed at the floor and eyes strained up, looking ahead. He stopped at his wife's side and rested one baseball mitt of a hand on her shoulder. "Need anything, love? Cold fruit?"

"No. Thank you."

Clay shuffled away.

"How long have you been married?"

"Forever."

There was white matter in the corners of her mouth and between some of her teeth. She swiped at it with a tissue, seemingly to clean it off. Some came away on the tissue; some remained on her mouth. Her body was small, about half my size, and wrapped in extraordinarily wrinkled skin.

"That's a long time," I said.

"We met when he was sixteen and I was eighteen," she said. "Got married when I was twenty-two. He was only twenty. We were babies."

Sitting beside her on the couch, my burning palms moved magnetically onto her, a phantom crackling sound accompanying their movement.

"I'm afraid."

"Tell me your fears."

I moved to kneel before her on the carpeted floor; I placed my hands on her legs just above her knees and she covered my hands with her own.

"I don't know what's gonna happen."

"When?"

"When I die. I think it's coming. And I'm ready. This old body. No regrets."

I nodded. My hands remained under hers in her lap. The heat there kept growing.

"What's gonna happen?"

This phenomenon of people looking to me as some sort of authority on death happened with my first hospice client, too. I wondered whether it was because I represented hospice, or because I practiced Reiki, or because death literally brought me to their door.

"I don't know, but I have some theories."

Her eyes were question marks.

"I think the first thing that happens when you eject from your body is relief, pleasure, bliss."

Her mouth dropped open and her eyes widened, as if I'd said what I wanted to say but didn't: it feels like a dozen or more simultaneous, sustained orgasms. I hesitated saying this next part, because I wasn't sure if it's what I believed, but it's what I sincerely hoped for, and I wanted to give her comfort in that moment.

"Then, I think we go somewhere else, and we see our loved ones. Who do you hope to see when you get there?"

She smiled and said, "My parents. Oh, I've missed them so much." Her wide blue eyes turned watery and dark like the sea. "I'll miss my husband," she said, and tears escaped.

"Oh, you know what else though? Time and space are different there, without a body. One day there is like one whole lifetime here. So you won't have to be without him for long. Just about five minutes, maybe ten after you get there, he'll join you."

"That would be nice. I want to see the world though, and what's going to happen. People are going to go to Mars in a spaceship and colonize space. I wish I could be there to see that. I wouldn't want to go into space though."

"Neither would I," I said. Here was a dimension of fear of death I hadn't considered: the fear of missing out. The fear of the world turning and time passing and humanity progressing without you as witness.

"But how about inner space, huh? You know, at the quantum scale, there's an entirely different physics than at the Newtonian scale, where we live. Particles, the smallest observable units of matter? They've been observed blinking in and out of existence. Where are they when they're not here?"

June blinked at me. Her mouth bent into a small smile. Her nose twitched. Tiny green fairies danced in her sky-blue eyes. "You must read a lot to think that way. I wish I would have spent more time reading."

"But do you see? Wherever they are when they're not here? Whatever space, whatever time? We're there and not there at the same time. Quantum physicists call that a superposition."

"Your hands are so warm. How do you do that? Have you always been this way?"

"I was told I had a lot of energy in my hands, and I should learn Reiki, so I did."

"It's true. Do you have kids? Are you married?"

"I have two kids. I'm divorced."

"Why are you divorced?"

Wondering whether to tell her the truth or skim the surface of the truth didn't take long. The connection we already had seemed as old as her marriage. I told her about my ex-wife, my gay husbands, our two kids.

"Well, that's OK," she said, smiling. "It's OK to be different."

I thanked her for the permission to be different and chuckled, moving my burning palms a few inches to a new, cool patch of skin. Her eyelids fluttered closed then opened. Her head tilted on her thin neck. "Are you tired? You should rest," I said and placed a pillow against her shoulder. She laid her head onto it and was asleep. I stayed for a while longer, then snuck away quietly.

///

"Has anyone ever told you that you look like the SlimFast girl?" Those are my words. My first words to you. A question that means: *you're so beautiful.* Because the woman who is featured in the television commercial for SlimFast is beautiful enough to be put in a television commercial for a product claiming to help turn women beautiful.

And even though you throw your head back in laughter and thank me, because, of course, you understand my roundabout compliment, my whole face turns hot and, I imagine, looks maroon from my embarrassment. Nevertheless, I hold my shoulders back and open my face into the best approximation of a confident smile I can achieve. This is the very first domino falling. Or was it before then when I watched you from the risers? Or before then, when I felt a funny feeling inside watching the woman in the SlimFast commercial? Or before then, when I dreamt of kissing women and touching their breasts while going through puberty? Or when I was born, an accident, the unplanned-for product of my mom dyeing her hair blonde and my dad getting too excited about that? (They already had four children they couldn't afford. Now, here was a fifth.) Or was it prior to that when I died before in another form? Maybe it is as old as the primordial ooze. It certainly feels that old, and big, and important. For a while, we're friends, neither of us understanding the electricity zigging and zagging erratically between us whenever we're near each other, so strong I can almost hear the static, smell the sizzle, see the smoke. After you graduate from high school, you don't go away to college but stay and go to the local college so as not to leave my side. I don't know it yet, but later, I will wonder if you stayed as much for your family as for me. Then, we kiss.

It seems as though I've written about it a thousand times, this first coming together of our mouths. But maybe those were just poetic recollections. We both love "Pachelbel's Canon" and play it on repeat on the boom box while hanging out in my room. Then the day arrives when we enter my room after darkness falls, extinguish the lights, and turn the volume on the boom box way up. Sightless on the floor, cross-legged and knee-to-knee, we sit. Slowly, our faces migrate together, closer and closer until our foreheads touch. Then, also slowly, our lips come together to touch, but just barely. That featherlight touch of your dry lips to mine, and the darkness and the breath — all my nerves turn to zapping live wires.

We repeat this ritual often, but never actually kiss. Our first kiss comes much later, when I bring you to Ohio State University to visit my brother and we get drunk together. There, on a couch in some stale-smelling frat house room, we finally kiss, our tongues entering each other's mouths with all the built-up hunger and force of roughly eighteen months of playing with the volcanic erotic energy between us.

Before that first kiss, there is a different frat house. There is the keg party.

It is the summer after your senior year of high school and before mine. You have a college ID already and at your orientation, you get invited to a keg party at a fraternity house.

"That's amazing! We have to go."

"You think? I don't think I can go."

"You're an adult! You're eighteen! Of course you can go!"

"My parents won't let me have the car."

"I have a car, we will take my car." After saving up an astonishing $800, I had bought my first car at sixteen: an orange Toyota Tercel hatchback. The pride and love I felt for that car was such that I stood on the back porch just to watch it sit there, shining in the driveway. `

In the basement of the frat house at the keg party, I am on my umpteenth large plastic cup of beer, and you haven't touched a drop. We sit on inflatable couches across from each other and college boys sit beside us, one with you and one with me. The one with me keeps talking, but I can't hear anything he says because I'm staring at the one with you. It starts with mild annoyance, and as he moves closer and says something that makes you smile, it blooms into fury. Part from simple jealousy, but part from what I know from experience drunk college boys will do, given the chance. I burst from my seat and charge over, grab you by the arm, and pull you up, drag you toward the exit.

"Let's go," I growl.

We ascend some stairs, and you're asking what's wrong. We burst out into the soft summer air of the night, and I pull you toward my car.

"Wait, you can't drive," you say. I reach into my pocket and give you the keys. We start toward the car when a male voice calls, "Wait, can I get your number?" We turn and it is the boy who was with you. He approaches and I charge him. The alcohol unleashes a new angry-dog energy inside me. With both hands, I shove him hard.

"Leave her the fuck alone."

He stares at me half-scared, half-amused. Just when I'm about to swing a fist at his face, you pull me away and toward the car. I stumble and start to fall, and you squeeze an arm around my waist to steady me. I hang on you and lay my head on your shoulder, stumbling toward the car, playing drunker than I am to ease my shame.

In the car, I continue to hang on you, stretching my torso across the small center divide, and looping my arms around your shoulders as you drive. The car is silent except for the whir of motion, the passing cars that sound like gently shoring waves. My face is close to yours and I can feel the hotness of my breath on your cheek, your neck, your ear. I look to your eyes and see pleasure surface there, your lips parting, teeth glinting in the oncoming headlights. Emboldened, I breathe more directly into your ear, kiss your neck, lick the slope of your shoulder and up the length of your neck to your ear, kissing and licking and nibbling, feeling the desire build in me with a force and ferocity I've not yet known, my body squirming with the heat, the urgency, the truth of it. This continues for the whole ride, the wet clicks and soft smacks of my mouth and the intermittent small pleasure noises from both our mouths the only sounds in the otherwise white-noise silence.

We reach your house too soon, and you pull over, park, turn off the car. You turn to look at me for the first time and I open my mouth to say something, but no words are there to meet me. What is unsaid and unsayable is there in my eyes, I can tell, because an expression of pain and fear falls over your perfect face and you turn and flee, jump from the car, run toward the dark of your house. Sobered by the intensity, I drive myself the short distance home.

The next day, we see each other at the high school, where we'd planned to meet at a volleyball game. We sit on the floor in the hallway outside the gym and marinate in awkward silence. Eventually, I look at you and smile and elbow your side.

"What happened last night? I was so drunk, I don't remember."

You look at me first with alarm, then you can see, and I can see, you don't believe me. You know that I remember, that I have endlessly revisited every moment, every touch. I look at my shoes and feel my face get hot. You grab my hand and squeeze it. What I think I read in your face then is something like pity, like you know I am desperately in love with you and you feel sorry for me because you would never, could never, reciprocate. A rush of anger animates my body and I shoot upright. Accusingly, I glare at you and huff a mocking burst of a laugh and say, "Wait, are you *attracted* to me?"

You cringe and drop my hand and say, "*No*!"

///

June and Clay rode a motorcycle for almost all of their seventy years together, never once crashing. I couldn't help but feel jealous. And even then, I knew there was more to my jealousy than the knowledge that they got to ride a motorcycle their entire lives without any consequence other than thrill and enjoyment. They met when they were kids. They were each other's first and only love. It forced me to remember the time when I thought that might be me: one half of a forever couple, endlessly devoted, those nights spent in the spate of kaleidoscopic love stretched over a lifetime.

"What's it like to be with someone for that long?"

She smiled. "Like having a home in another person."

I shifted my hands a few inches up her legs to spread the warmth.

"That's amazing! Your hands are so warm! How do you do that?"

I felt the tense muscles of her legs drinking in the warmth, like a dry cloth slowly softening and opening as it absorbs liquid. She'd been experiencing cramps that were painful enough to wake her from sleep. Under my hands was what felt like a twitching knot of taut cords.

"It's just energy. Really, it's just love."

I'd been trained and attuned a Reiki energy healing practitioner in my twenties, my early sobriety. I went through level one, two, and then master teacher training. I'd never used it to make money. It had always been an offering, an act of service. When signing up as a hospice volunteer, I did so in response to a call for volunteers who practice a form of energy healing.

I moved my hands around her waist to the small of her back, bumping into the oxygen tube snaking up her shirt. It split in two behind her head then came over her ears to loop under her nose. The machine hissed and gasped at my feet. My arms circled her awkwardly, but I sensed her back needed it, so I suggested she lie back on the couch. Next, she was in my lap, and I held her rather like a baby. I slid one hand under her back and kept the other on her chest. She fell asleep. I stayed that way for as long as I could while my legs, bent under her, fell asleep. Somehow, I was able to slide out and quietly leave without waking her, staggering to my car on wobbly legs.

///

This tyrant we call time. In the end, it only gives us two-and-a-half years together. The tail end of which is spent arguing and crying, feeling blue.

"We can run away, where your parents will never find us. We can be together," I implore you. Your parents are Greek immigrants, very religious, and very homophobic. They have sensed the bond between us and have punished you for it, tried to keep us apart. We went to great lengths to be together anyway, until now. You're slipping away. But not because you don't love me. Because you're afraid.

"But what if my parents are right? And this is wrong. And we're sinning."

And then I say it. The words that so many songs and movies and novels have been made about. The words spoken so many times by so many humans throughout history, they've become cliché. "But how could something that feels so right ever be wrong?"

The problem is, we're both right. And neither of us is wrong. For the next few decades, whenever I see or hear that Rumi quote — *Out beyond ideas of wrongdoing and rightdoing, there is a field. I'll meet you there* — I think of you. Our love existed in that field. It exists in that field still. I imagine the field elevated and green and flowered and dancing with mist.

But before the end, our first nude scene. I act experienced, even though I'm as new to all of it as you are. It's all burning breath and pressing wet and both of us together inside. It feels like waves of your body crashing over me.

Your first year of college, you miss so many classes that you fail every one. My last year of high school, I miss so much school, I almost don't graduate. We don't have time for school. We don't have enough time, ever. Touching you becomes like peeling an orange — impossible to do without getting my hands wet. Each separation is a filling up to brimming, then spilling over with longing. I come back to you each time overfull with desire, waiting to be emptied. But I am never emptied. A whole ocean of appetite lives inside me, teeming, roiling, inexhaustible.

We find spaces in which to make love. The library after it closes. The projector room in the auditorium of our college. The couch backstage in that same auditorium. Once, on that couch, when we're naked in each other's arms, someone comes backstage. Heavy footfalls first, then a deep voice saying, "Hello?" I throw myself on top of you, covering you. The footfalls halt behind me, silenced for a beat, then retreat. For the rest of our time together, only about a year, you cite that moment as one where I was your valiant and brave hero. Shielding your nakedness with my own, you say, had revealed so much about how deeply I love you.

And it's true. How easily I could have spent seventy years with you, loving you that way. How easily you could have done the same with me, if only I had the anatomy deemed acceptable to your parents, your religion, your culture, your fear.

Nevertheless, you are accepting of me, even with all my faults. You don't like that I drink too much, but you love me anyway. You are kind and intelligent, gentle and tender. While I go first to the power and hardness of anger, you are more prone to the real and vulnerable sadness underneath. With you, I feel safe to allow the sadness. And when I collapse into your arms and cry, you hold me. There, in that silent belonging with our arms wrapped around each other and my face in your neck, is where I build my child heart's home.

There is no proper ending. I keep asking for one last time, one last time. And you keep showing up. I think we might make forever out of one last times. But then one time, you just don't show up again. You quietly remove yourself when I'm not looking. While I'm on the way.

And yet, we had passed lifetimes together in a single afternoon, in our bright field of mist, skin to skin, mouth to mouth, mouth to skin. Those two-and-a-half years, they are two-and-a-half days, and they are seventy years, both.

/ / /

"I missed you," June said each time I came back, taking my hands in hers and pulling me close.

"I've been seeing things," she said on my second to last visit, "and I've been sleeping a lot. I'm so tired, I just want to sleep and sleep."

"That's normal and it means you're getting closer. What have you been seeing?"

"Oh, it's there in the corner of my eye, a bouquet of flowers, blue and red and orange. And above that, a light. A green light."

"Does that mean anything to you? Did your parents ever give you a bouquet of flowers?"

"I'm sure they did. I don't remember."

"Maybe it's a greeting? A welcoming from the other side."

"I'm scared," she said. "How will I reach you? When it's time? I want you to be here."

"Your husband and your daughter have my number. They can call me. I'll come as fast as I can. I'd be honored to be here with you for that."

At my last visit, she'd been to see the eye doctor. She'd told her daughter about the flowers and the green light, and in response, her family took her to get her eyes examined. It made me a little amused and also a little angry.

"I have perfect vision," she said.

"Of course you do. There's nothing wrong with your eyes. Or with any part of you."

"This old body. I miss how it used to be. I'm worried about my sons. I'm worried about Clay."

"Why are you worried about your sons?"

She told me how they argue, how they hate each other, how they can't stand to be in a room together. One of them, she said, loves Trump and won't get vaccinated. The other is more like her, and he believes in science and thinks Trump is horrible. I held her in my arms until she fell asleep, which had become our routine. She wanted to be held, and I wanted to hold her. In between my visits, I felt a longing to hold her again, and I came to her overfull with that desire, waiting to be emptied. Her body felt so hollow, so light, and yet there was so much heaviness too. The heaviest heaviness.

After she fell asleep, I went through the kitchen that smelled of hot dogs and oranges, through the sterile dining room with its unused dining chairs and a plastic-covered table, through the sliding glass door to the patio, where their old dog growled at me viciously, protective of his papa, to talk to Clay. He was seated in a chair at the round table.

"Be quiet, get back," Clay shouted. The dog's ears flattened against his head, and he obediently retreated. I asked Clay about their sons.

"June told me she's worried about them. She wants them to get along before she dies," I said. He chuckled and shook his head.

"That'll never happen. They can't stand each other," he said.

"Do they love their mom?"

He nodded.

"Can you ask them to pretend? For her sake? So she can let go of that worry and die in peace? Except for that, I think she's ready."

He looked at me thoughtfully and nodded again. "That might work," he said. "I'll ask."

"She's worried about you, too."

His eyes turned red and watery. He blinked back tears. He took his large hands from his lap and put them on the glass patio table. It rattled under their weight.

"Are you ready?"

"I'll never be ready." His voice trembled.

The phone call came on a Saturday evening. June's daughter told me she was close. She was having a lot of trouble breathing and they moved her from her bed to a hospital bed out in the living room. I agreed to come the next morning and every day until she passed, secretly willing to call out of my paying job if necessary, in hopes I would be there with her, which was her wish.

The next morning, I went to the farmer's market on my way to June's house to buy her a bouquet of flowers. The red and the orange were easy. It was a challenge to find blue ones, but the florist brought them out when I asked — Love in a Mist.

"That's what they're called?"

"I know, isn't it beautiful?"

I blinked back tears and swallowed over the lump rising in my chest. Walking out with the flowers clutched in one hand, I held my phone with the other, and a text came through from June's daughter. *Don't come. Mom's gone.*

I stopped walking and dropped the phone. I gripped the flowers harder and pressed them against my chest. My mouth fell open, and my heart pounded. I closed my eyes.

Goodbye, sweet June. I miss you. I love you.

The flowers lived too long, cut off from their source as they were. I kept them beside my bed, where they received no natural light. There was no explanation for their unapologetic aliveness, blossoming for weeks. I wanted to see them out of the corner of my eye as I lay down each night to sleep, and each morning when I woke. Once or twice, I even imagined I saw that green light.

Whenever I thought of Clay, I thought of you, my long-lost first love. After it was over between us, you married the nice Greek boy from your church, the one your parents approved of. You stayed in our hometown and had three children. Your children went to school with my sister's children. Somehow, I swear, in some quantum alternate universe, you and I were June and Clay, and those children were our children, and maybe those flowers were from that universe. Maybe that Love in a Mist is ours. Undying, even in the dark.

La Petite Mort

My daughter was newly two years old when I first met her. Still in diapers, but with a large vocabulary, she was a ball of joy with a wide-open heart. Among the first dozen words she spoke to me was "I love you." Her mother was my new girlfriend. I'd left my previous girlfriend chiefly because she was certain she didn't ever want children, and I was, to my surprise, certain I did. At thirty-four, I was afflicted with all the attending clichés about a woman of that age. Suddenly, I was aware of the eggs I had been carrying inside me since birth, and aware of the colossal potential each of these magical single cells contained. I learned every woman is born with all the eggs she will ever have, approximately one million, and approximately 70% of those will die off before she ever ovulates for the first time. Once menstruation commences, one egg each month will descend slowly from an ovary through the fallopian tube and into the uterus. This lining of the uterus makes a nest, like the waiting nest of birds, and it thickens in preparation. Hormones get released, the effect of which is the woman feeling desire for sexual intercourse. Meanwhile, that stockpile of eggs dies off in droves. Of the few hundred-thousand that remain, a mere 300 to 400 will be ovulated in a woman's reproductive lifetime.

Each of these eggs producing a completely distinct person was the first idea to blow open my mind, expand my thinking into new vistas. Suddenly, I yearned for the potential human one of these eggs would produce, and was excited to meet them and simultaneously grief-stricken about the other hundreds who would never breathe air. When I started trying to get pregnant, I learned about how unlikely the conception event is. So unlikely, in fact, I walked around in jealousy and wonder seeing pregnant women everywhere. It's the most commonplace and impossible miracle among womenkind.

More than a decade elapsed since the crushing loss of first love, more alcoholic drinking, the beginnings of my scientific training entwined with my energy healing training. I was pre-med as an undergraduate, but I was also an active alcoholic and a child. After four years of college, I looked at my transcripts and saw that I had accidentally taken enough literature and writing classes to get a degree in English but had another full year of prerequisites for medical school to complete. I opted to be done and got my bachelor's with a dual major in English and Biology and a minor in Chemistry. My first job out of college was as a copywriter for a healthcare communications agency. It would be many years of honing my scientific expertise through heavy reading of published research in both oncology and neurosciences before I would get my master's in Science Writing and then another in Regulatory Sciences, and emerge as a clinical science consultant, helping to bring effective treatments to the market. And through those years, several more loves came and went.

After joining an online community of queer women trying to get pregnant, center stage among the glut of tips and tricks offered was an orgasm will help your chances of conceiving. This sounded intuitively right to me. In fact, when I first started having orgasms, around the time of puberty, I remembered imagining an egg popping out into my uterus with each one, complete with a *Ding!* like in a pinball machine. The event was so epic for my body, it logically followed it would play a central role in conception, as does the male orgasm. Curious to have that childhood fantasy confirmed, I looked for research to back it up. What I found left me skeptical that science — or, at least, what's available among published research — has the whole picture of the role of the female orgasm in the mechanisms of reproduction. Because the only thing proven in the lab up to now is oxytocin is the hormone responsible for pelvic floor contractions during orgasm, and some studies found experimentally administering oxytocin to women increased the amount of sperm-like fluid transported to the fallopian tubes — meaning, since orgasm increases oxytocin, orgasm could then make sperm transport faster and conception more likely. And though it did seem to me that there might be a biological role of the female orgasm in conception, this idea also made me think about the girls and women who are raped, trafficked, married off against their will, and I don't want to imply that if they got pregnant, they must've enjoyed it. There was a man — a vile old white anti-abortion Senate candidate — who, in keeping with their tradition of ruining everything, claimed that women rarely get pregnant from "legitimate rape" because they can somehow shut down conception. No.

Happily, though, I added an orgasm to my arsenal of details making up what felt like a ritualized practice of daily attempts at conception each month for five days starting with the day of ovulation. My partner at the time, my daughter's mother, was involved with the first attempt in a way that was both comic and tragic. What we learned from that experiment is insemination is not sexy, and we emerged from it with her as the first and last partner to ever peer up close at my cervix by way of a headlamp and speculum.

///

The early Greek philosopher kings had entire cosmologies thought up — whole internally consistent systems of thought that include an organized explanation for everything that is. Aristotle's cosmology was based on the observable universe, which back then meant what the unaided eye could observe. He therefore theorized the Earth was the center of the universe, orbited by the sun and other celestial bodies. And humans were the most important part of the universe in his estimation. He also surmised the motion of heavenly bodies is circular, and this pattern, having no beginning and no end, evidences an aspect of the human animal that has neither beginning nor end but is eternal.

Plato laid out his cosmology in the book, *Timaeus*, which includes his ideas about the origin and nature of everything. Like most ancient Greeks, Plato was a polytheist, and his position was that this imperfect yet highly intelligent world in which we exist was created by a Demiurge (from the Greek demos or people and ourgos or work) – a superior and benevolent god, who strove to create an image of itself that was as good as possible. But the Demiurge could not create a world out of nothing; its powers were more limited than the god of Genesis. The Demiurge fashioned the cosmos out of materials provided by a pre-existing chaos, or jumble of matter, which it organized into the four elements – Earth, Water, Air, and Fire. These formed the body of everything. The soul of everything, which Plato considered as superior, was its principle of eternal roundness, bringing about the circular motion of the moon, planets, sun, and stars.

What I love about it (and about Plato, in general) is it includes a mathematical precision on one hand, where he assigns tiny geometrical shapes for the body of everything, these elemental phenomena, and demonstrates how they can combine to turn one element into another and how they also create what we experience as sense perception (fire gives way to sight and water, hearing, etc.). On the other hand, he blurts the most outrageous mystical assertions for other things, like: "And then god lit a torch and that became the sun." I have to smile. Because the limits of reason and logic are like the horizon on a cloudy day, where water blurs together with sky as though in open conversation. So too does the logical blur together with the mystical when they, together, speak.

Like Plato, it is toward the mystical I lean to get an understanding of what I'm doing here, and what forces are at play that make me do what I do.

///

The other variable in this baby-making equation was the boyfriend of my partner's ex-husband. We had been co-parenting our daughter for two years together, and I admired them as fathers. It was my girlfriend's ex-husband who had the idea his boyfriend and her girlfriend (me) could make a second baby to grow the family. Having been on fire to reproduce for all those two years and then some, I agreed it was a brilliant idea. The whole operation was DIY for the first five months. At about five consecutive days per month of attempts, I had inseminated myself about twenty times over those first four months and not gotten pregnant. By the time the fifth month approached, we were planning to enlist the help of a clinic and get intra-uterine insemination, or IUI, which improved the chances from 10% to about 15%. But it required having the semen "washed," which isolated the sperm and used only the most virile among them. The state of Massachusetts, where we were living, had a strange and seemingly homophobic law in place about only allowing known donors if that person was the significant other of the woman. In other words, I couldn't use anyone's semen unless I was fucking them. This confused me, as I had voluntarily had this man's semen in my vagina dozens of times already, so how was that so different from their requirement? Regardless, the law was the law, so we were prepared to go in and impersonate a couple as a work-around so we could get the help. But we never had to compromise and potentially traumatize ourselves in that way, because our attempts in the fifth month were fruitful. And the orgasm I had to help create our son was memorable.

/ / /

In Greek mythology, Eros is the god of love, a winged male creature who is the offspring of Aphrodite and Ares. Aphrodite is the goddess of love and beauty, and Ares is the god of war. Eros inherited these combined powers, giving him the ability to make mortals and gods alike fall in love. The progenitor of Cupid, he achieves this by shooting them with arrows.

Then there is Thanatos, the god of death, produced by Nyx and Erebus, the divine parents responsible for darkness and night. Thanatos has a twin brother, Hypnos, the god of sleep. Thanatos is associated with violent death and is called upon by gods who wish to inflict death upon one another. Thanatos often manifests anger, violence, and aggression. These characteristics are portrayed in those who are driven by death instincts.

/ / /

I once dreamt that my house was broken into. In waking life, I had just been dumped and was broken hearted. The dream happened while I was away on a spiritual retreat and nursing my pain. In it, I came home and heard the rustle of an intruder in my bedroom. I ran toward it and there was a figure by the window, then just outside, then far away. The curtains flapped in the breeze.

I turned around and saw a large cup of change I kept had been dumped out on the bed and was spread all over. At first, I assumed it was because I'd been robbed. The intruder had taken the big change and other valuables. But as I looked around the room and took stock, nothing was missing. In fact, not even the change looked to be missing.

Instead, it had been organized. All the quarters were neatly stacked, the nickels, the dimes.

A Jungian analyst told me the dream was a beautiful metaphor reminding me when I feel taken advantage of and vulnerable, it's because change moved through. But it didn't diminish my value. It only reorganized it. I loved that idea then and I love it now.

So, when I arrived in Ohio with my eighteen-month-old son and saw some of my family who I hadn't seen since my wedding a short seven months earlier, I felt some discomfort. But I also know the vulnerability I felt from having already filed for a divorce wasn't what it seemed. It wasn't a diminishing of my value. It didn't mean I was less than my siblings who had all been married to the same person for double-digit years.

/ / /

Rather than Sigmond's famed yet reductive interpretation of these Greek gods, Eros and Thanatos, as the two fundamental drives human animals can experience — either the drive for life (sex and reproduction), or the drive for death (destruction) — I would flip the second half of this equation and remove its either/or. I believe sex and death are the biological sources of all human ambition, and both are always at play simultaneously. There is the drive toward sex and the drive away from death, or, in some cases, toward death, when transcendence is sought.

When I first connected sex and death together, perhaps like many, it was because of the French phrase for orgasm, *la petite mort*, which in English, translates to "the little death." Georges Bataille was a French philosopher who wrote about eroticism and death. Specifically, he likens the state of mind one reaches during orgasm to a loss or semi-loss of consciousness, a little death, and therefore coined the term in French. In my own experience, I have never felt a loss of consciousness or otherwise death-like state during an orgasm. In fact, I have often had experiences during orgasm I can only describe as a heightened, elevated states of consciousness. I've seen sounds, smelled colors, felt scents against my skin. Once, in the early morning, I had two sequential orgasms during a dream, and woke after the second with a vivid understanding of a dimension of human relationships I called the z-axis.

Like the geometry of time and space, with width, height, and depth, human relationships also have an x-axis, a y-axis, and a z-axis. Along the x-axis, is the couple's projection into the world. How widely is their association known? How big of an audience does it have? Since the advent of social media, I have been attracted to women who are shy with this x-axis, by and large. None have had a desire to broadcast our relationship on social media, although a few have allowed me to do that. And secretly, I wanted to have a partner who was so proud to be with me she would want to expand that x-axis, yet with taste and appropriate restraint. The y-axis is the highs and lows a couple can reach. How much fun they have together, and how miserable they can make each other. I've always had plenty of range on the y-axis with romantic partners. The z-axis is the erotic dimension. Like a body of water, it has varied depths. And, like a body of water, its depths are hidden. You may be able to see the surface and sense when it's deep, but you can't know for certain without diving in.

///

I was in Ohio for a couple of days to help my siblings clean out the house we all grew up in. My mom had lived there for nearly fifty years. She lived there on her own first, then with her first husband, then again on her own with three small children, then with my dad and two more of us. Fast forward thirty-eight years and there we all were, helping her to face the biggest change she had ever had to face. I helped dig through a single closet in a single bedroom. It took hours to unpack one packed corner of one closet. We found untold treasures: old report cards, our first drawings, amazing photos like all five of us lined up from tallest to shortest dressed in underoos — underwear versions of superhero outfits. We found untold horrors: The red and blue windbreaker my sister was wearing when she had her car accident as a new driver at age sixteen — the one they cut from her unconscious and broken body. It happened four days before my eleventh birthday, on May 9, 1988. We also found the first shaky writings she did while still in a coma about five months later that read: "TAKE ME HOME." We found a thick stack of paper that was a report of all her injuries, including to the pelvis, the brain, the neck, spine, knees, ankles.

At the time we found these items, I was less than two years away from experiencing my own near-fatal accident, one that broke my bones and severed my hand but left my brain intact. My sister hadn't been so fortunate. Her head had smashed into the windshield, which bounced it like a basketball back into the front seat, which she rolled off into the wheel wells, wherein the car closed on top of her, trapping her in. Rescue workers used a very large crane-like structure called the Jaws of Life to tear the top of the car off and retrieve what they were sure would have been her lifeless body. But she was, albeit only barely, alive. And though some doctors predicted she would not live long, or that if she did, she would remain in an unresponsive state, she did regain consciousness and survive. But she would have a long road ahead, would never return fully to the person she was before, and would require assistance for the balance of her life.

At age thirty-eight, I sat suspended between my traumatic past and traumatic future, looking through artifacts that reminded me just how fragile we humans always are. The letters, the boxes, the old clothing, the change, all of history — it surrounded me.

///

Perhaps for men, there is a death-like state associated with sexual climax. Contrarily for women, as borne out in a 2017 study on brain activity in women during orgasm, measured via fMRI, extensive cortical, subcortical, and brainstem regions reach peak levels of activity at orgasm. This likening of orgasm to death is apt, but not for the reason Bataille theorized.

As during the female orgasm, in the moments preceding and possibly the very moment of death, brain activity climaxes. Though the moment of death remains hotly debated among neuroscientists, whether it's defined by cardiac arrest or brain death, all brain activity is thought to cease by around three to four minutes from the moment the heart stops. And in those final minutes, the pattern measured on an fMRI looks similar to that during the female orgasm. The natural conclusion I come to by way of this evidence is peak brain activity, and death, as during orgasm, is ecstatic.

There is an additional, ultimate relationship sex and death share, and that is how, in my life at least, romantic relationships, where orgasms are produced, die. It took me a while to understand romantic love is inherently conditional. There are always conditions for this most intimate of intimacies — behaviors and attitudes both partners must, more or less, sustain and maintain for the relationship to continue in a fulfilling way. They often include such behaviors as monogamy, thoughtful gestures, remembering birthdays and anniversaries, and, of course, ongoing sexual desire as well as the regular performance of the act itself. Attitudes required typically include respect, willingness, interest, and honesty. When these behaviors and attitudes fall to the wayside for too long, the relationship will end. And in my experience, they always fall to the wayside. My longest romantic relationships to date lasted only about four years each, back-to-back, and recently, at my most mature. And I believe that is because I am not a person to tolerate the loss of these fundamental behaviors and attitudes that keep a romantic relationship feeling vital and fulfilling.

///

I found a couple of letters written between my parents from when we were all little, arguing about the same issue that ended their marriage decades later. And strangely, this comforted me. It made me feel relief my ex and I had the insight to end our marriage then, despite its newness and the stigma, and save ourselves decades of forcing our relationship to be what other people expected it to be. Even long after that phase of it had ended.

A wise friend said something startling while I was feeling some unnecessary shame about it. She said, "I don't care if my last relationship is three months old, I'm going to die happy." And even though I know a significant partner's function isn't to deliver my happiness, I also know my happiness is the first thing to get sacrificed when I am not being true to myself. When I am not walking and living in my truth, despite the judgments that will inevitably come.

///

My most recent long-term romantic relationship was with a woman who had first been my best friend for four years. And when I proposed to her we convert into lovers, her only objection was spoken with her characteristic lucid eloquence: "Romantic relationships end, and I can't lose you." In response, I promised if it didn't work out, I would go back to being her best friend, if she would do the same. She agreed, but in the end, she did not fulfill that agreement. And the grief over the death of that relationship lingers in me still, years later.

Even if a relationship lasts, it changes. And perhaps that is all death ultimately amounts to. Not an ending. A change. And with each significant change comes loss. A little death. La petite mort. Life is a series of little deaths all in a row.

Since it seems true romantic relationships do inevitably end, it's a wonder I keep going back for more. After my last long-term, I decided monogamy was the problem. So, I tried non-monogamy, and that experiment ended badly. Among its many problems was the risk involved of having sex with people who were having sex with other people, and at this point in my life, I don't have that kind of appetite for risk. And one main ingredient in my desire to keep trying monogamy is my desire to have a robust sex life, as I'm acutely aware of my impermanence, or rather, of the impermanence of this body I inhabit. I will be a sexual being for a finite number of years, and I'm ambitious with my goals for those years. Maybe my optimism is stubborn, and through those rosy glasses, I can see a version of monogamy that will work for me. In it, both individuals feel freer than either of them could on their own because they support each other and lift each other up and believe deeply in each other's talents and pursuits. They are interdependent, holding hands and walking toward a scary, shared goal. And they are autonomous, with full, separate lives they can come home to each other from, ripe with stories to share. It's monogamy with healthy boundaries, boundaries which define the essence of each person without rigidly cutting off connection. Porous, vigorous boundaries. It's a vision of monogamy I think is possible, if only I can find a like-minded woman who wants roughly the same things out of a relationship, and with whom I have ample depths to plumb along the z-axis.

As a woman who loves women, I have what my experience has proven to be a higher than average libido. Almost all my partners have not matched me there. And although I've always been afraid of death in theory, in that I'm afraid of loss and of the unknown, the time came where I faced the real possibility of death — in a catastrophic motorcycle accident — and I panicked. My response to seeing my hand hanging off by skin, splintered bone and blood in its place, and feeling myself bleeding out through my femoral artery was to have the first of many panic attacks in my life.

Two bothersome personal truths drove me to interrogate these topics closely: the trouble I've gone through in service to my desire for sex, and the fact that I was going to meet the inevitable end for us all while completely freaking out. I stubbornly believe the quality of my death somehow matters, and so I'd like to have peace in my mind the next time I face it. To achieve that, I believed I needed to overcome my fear of death. In my experience, the best way to stop being afraid of what scares me is to move closer to it. Befriend it. Get intimate. Thus, volunteering for hospice.

And, as for sex, I'm interested in what it has to teach me. I'm interested in that ecstatic bright brain, ablaze from the little death, and in what it affords.

///

At the end of the day, late into the evening, we were still upstairs going through everything. My oldest sister unfolded a piece of paper, read a few lines, dropped her hands down, sighed, and said, "This house is so heavy." I took the paper and wrote on it, *Change makes cents,* and handed it back to her. She read it, she got it, and we laughed.

It's true. That house was dense with artifacts that chronicled five decades of life lived within its walls. All we came across that day had barely scratched the surface.

And the ubiquitous item being excavated, among everything else: change. Lots and lots of cents. I had to rescue more coins from nearly going into my son's mouth than I could count. It filled my pockets. Filled jars. And the metaphor was not lost on me. When I say change makes cents, it's not only a pun but has the purpose of illuminating the inherent value that shakes out of every crack in every surface getting broken up by the rumble and storm of great change moving through. All of that change, all of the excavation, was bleeding with value.

THE HOUSE OF SEEKING

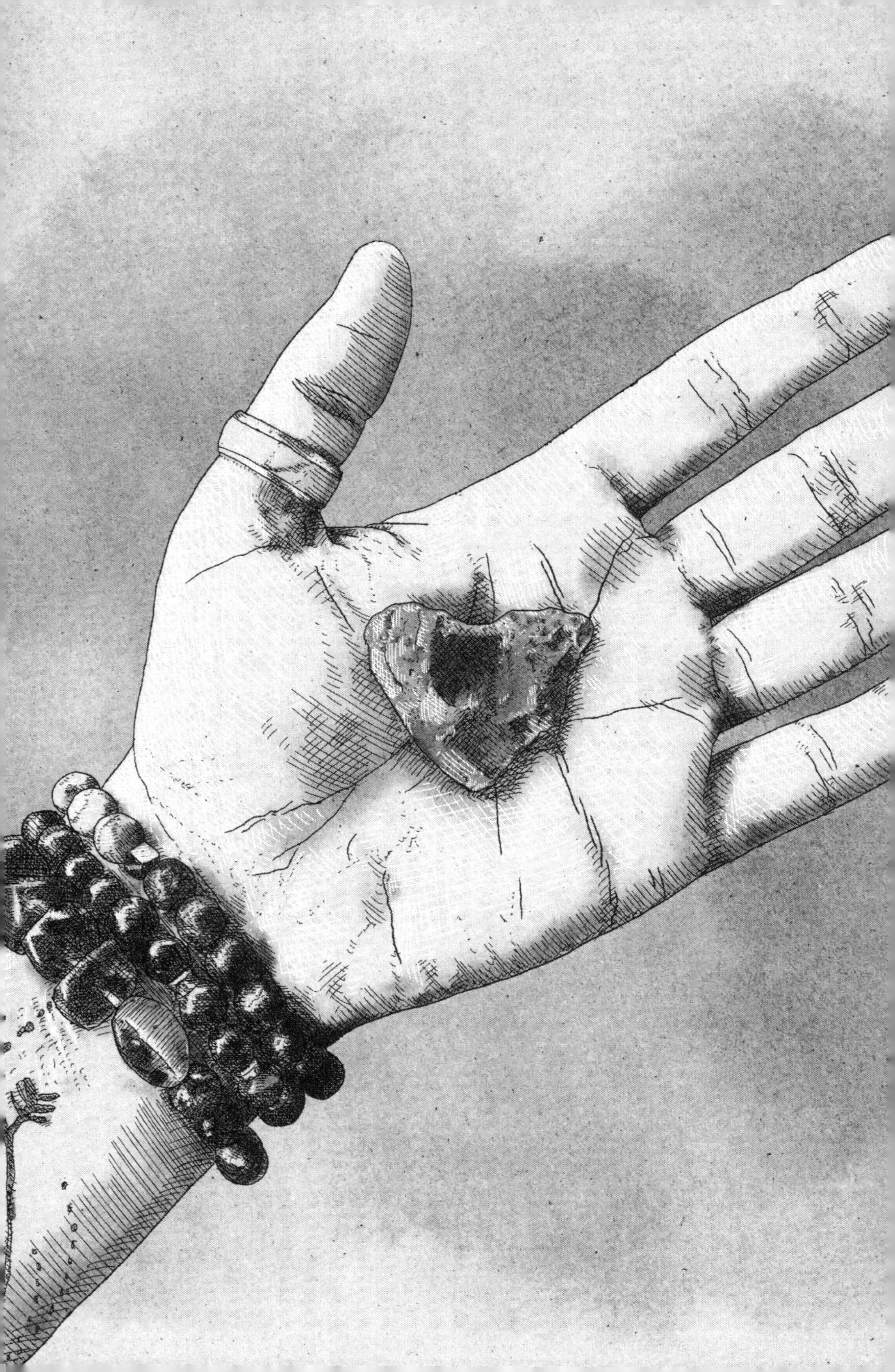

El Corazón

It's been so long since I wrote about you, my sixth love, and about our love affair. Use of the word, affair, here is deliberate and emphasized for its double meaning. This is the word often used colloquially for illicit, extramarital romance. This is the word I heard repeated on the phone in a man's voice. That disembodied voice sometimes briefly droned in the background during our long phone conversations. I remember hearing it and feeling simultaneously curious — who is this man you share a home with? — and annoyed — who is this man rudely intruding upon my fantasy?

But the time I'm referring to, when I heard the voice repeating the word, was more than a decade ago. It was the first day of a new year. To my surprise and delight, and rather last minute, you asked if you could spend New Year's Eve with me, just the two of us at my house. I attended an AA meeting and you texted me the question. Whatever plans I might have had for after the meeting dematerialized right then. Of course, I responded and followed with, *I can't wait to fuck you senseless.* You let me know, in response, you dislike that word. *I don't fuck*, you texted. I apologized, and explained I simply like the word, specifically when it's used to refer to an act of sexual congress. Its brevity as well as its sounds, opening soft and finishing hard — they seem to match the act, especially when it's well-executed, as it consistently was between the two of us.

That night, we fucked for the second or third time in the shower, then left ready to do it again. I lay back on my bed and you crawled on top of me. Your hair was still wrapped up in a towel heaped atop your head. I remember your face and the look in your eyes — playful and hungry, your gaze flitted from my eyes to my mouth, and you moved in to kiss me, then your phone rang. It was well past midnight at this point, and you jumped to answer it. I didn't move.

"Hello?" You sat up in bed in half light, a towel loosely wrapped around your body and the other still atop your head. I lay naked right beside you, waiting.

"Are you having an affair?"

It was that disembodied voice, but the clearest I'd ever heard it. And the clarity, the closeness of the voice, made it fill in with a body in my imagination. On the other end of the phone was a lonely and tortured man, knowing the truth, but unable to face it. Here he was, asking you to help him face it.

"No," you said.

"Are you having an affair?" he repeated.

"No," you repeated.

"Are you having an affair?"

Again, you lied. In my memory, he asked the question a fourth time. Each time, the voice grew a bit louder, a bit more desperate. The voice liked the answer, but also knew it was a lie, and this turned up the volume, hued the tone with desperation.

"I'm coming home," you said, simply and finally, then hung up. You stood up and got dressed without saying a word to me. I didn't move. I didn't speak. My heart was beating so hard, I could have sworn you heard it, too, and tears stung my eyes, dousing the fire that had just been burning inside them, the way they seemed to burn when filled with your face, your body.

The way I remember it may not be accurate, and it's one memory I didn't have the guts to bring up with you when we were together. We exchanged a few recollections, and none were shared. This new incarnation of our connection does vibrate with the same, original energy, but we're older now, and wiser. When I sent you a text message on your birthday, I didn't expect a reply. After sending a few messages over the years, some with requests to see you when I was going to be coming through, you either didn't respond, or politely declined. But this time, you replied, and a lively correspondence ensued, culminating with a three-hour phone conversation. Daily interactions and a platonic, unconditional love have been what we've shared ever since.

And yet.

The love conceived of by Plato, although of a non-attached variety, was not devoid of sexual energy. In fact, the young men who were students of Plato did engage with him sexually, but because there was no anal penetration involved, he claimed the love shared among them to be merely friendly. Instead of in the anus, they pumped their penises between one another's tightly closed thighs. And those friendships with acted upon erotic energy, and with sexual encounters, were deemed platonic.

After the phone call that shattered the first morning of the New Year, we didn't talk for a while. I didn't hear from you at all the whole next day. Alone, that evening on January 1, I went to the movie theater and saw the film adaptation of Cormac McCarthy's post-apocalyptic novel, *The Road*. It was so dark and depressing, when I walked back out to my car in the cool desert night air, I had to sit on a concrete parking curb, burdened as my body was with the heaviness of how it ended, both the movie and our affair.

But our affair hadn't ended. Not yet.

After that, it did end, but not right away. We kept coming back to each other, even after I moved away and lived in another state. When you were finally gone, your absence was so big for so long, it was almost more of a presence than your presence had been. Your name tumbled around inside my body, between my legs, between my ears, often out my mouth. My heart turned into a graveyard where I laid to rest all the memories, all the pleasures, and all the pains. I planted your name in my graveyard heart, too. I never expected anything good to grow there.

But then, over a decade later, you re-entered my life. And now, you're back, here in my life. Your presence can't be anything small, I see that now. Nor can mine for you. And because we enjoy each other so much, we decided to take this unique connection we have and call it a friendship. We can be best friends, and love each other unconditionally, and have each other always. But not all ways. Romantic love is inherently conditional. And because those conditions tend, at least in my experience, not to be met for long, romantic love has always ended. Ours did, and those since ours also did. Each one a little death in succession.

When you came back into my life, I was telling myself a story again. This lovely narrative I spun had me placed just at the beginning of an important transitional phase. The great alchemist in the sky was preparing to do her transformational work with me. Of course, you would be a part of that. How poetic.

She would be changing me from someone who is fundamentally unavailable for a long-term, committed, romantic relationship (and so who therefore necessarily attracted others who were also fundamentally unavailable) into someone who is available for just that, or so I planned for my story's conclusion.

I wanted that to be the point of all the change and growth here, around the midpoint of my life. Even if becoming available meant I got the version of it I most feared: one that settles into monotonous routine, drained of its passion, resembling more a sibling relationship than a romantic one. I declared to be available for all of it, come what may. I thought I already had, and the universe had given me exactly what was needed to facilitate that change for me.

Part of it could have been the meditations I did by candlelight with crystals, when I visualized my heart as open and receptive to love. Or, more specifically, committed, romantic love. Part of it could have been the rituals involving sweat, sunshine, and salt water, allowing the great mother ocean to wash away all barriers to this variety of love. And also, and of course, there was the continuation of the phenomenon I had been experiencing for many years by that point — of finding heart-shaped rocks everywhere in nature. Always, everywhere I went, I saw them. Often, I picked them up and saved them, collected them, gave them away as small gifts, as tokens of my love. Their metaphysical significance, I decided, was to remind me love is abundant, and everywhere, and available always in many shapes and sizes. One of them I kept, had made into a pendant, and wore as a necklace. It became a powerful talisman of self-love never parting with my person. All of this contributed to my healing toward wholeness and true, open-hearted availability.

But the thing I believe facilitated my transformation to completion was an unexpected and powerful connection I had when visiting my other best friend in Mexico.

Heather is full of joy and optimism. Her bright light and warmth attract people to her, and she makes friends easily with almost everyone. And so, I wasn't surprised when I came to visit her in her new home in San Miguel de Allende and found she already had a host of friends, waving and calling out "Hola!" to what seemed like every other person we passed on the street. This was so frequent, in fact, I grew a little impatient with it, wanting to move along to our destination with more expediency, versus stopping to converse with each one of the people she knew. When we walked past the open door of a storefront and she shouted greetings into it, and stopped to say more, I kept walking, hoping she would keep it short and hurry to catch up.

"El, come back, I want you to meet my friend. She's a yogini like us, a kundalini person."

I put a smile on my face and retraced my steps to the open storefront where she had stopped. Inside was a little boy with big brown eyes and a mass of curly brown hair, staring at me. When I saw him, and he saw me, our eyes connected and did not break contact. He spoke something in Spanish.

"He's interested in your necklace," his mom said. "Is it a heart-shaped rock? He finds heart-shaped rocks everywhere. They just appear for him out in the world."

When I heard this, my mouth fell open. Instantly, the inexplicable draw I felt to this small person was explainable. Here was another person for whom this phenomenon of finding heart-shaped rocks everywhere was a notable and defining feature by the people who knew and loved him.

I approached him, took his face in my hands, and told him we were soulmates. I wanted to tell him, te amo. I love you. But I held back that much, as there were spectators, and I didn't want to seem creepy. I was also conscious of myself as a white woman from the US posing with a child in another country and being inappropriately affectionate. I thought of pictures of young blonde women posing with a crowd of African children on their spring break mission trips. Jumping to telling this young child that I love him felt a bit overly intimate in that way that could be racialized and uncomfortable. I learned his name, Dimas, and I exchanged contact information with his mom, Gab — a tall woman with a striking face and short, dark hair. That first day, I was too shy to tell Dimas I loved him.

But the next day, I told him. Over and over, unabashedly. I brought him two heart-shaped rocks I had found, one small and one a bit larger.

"Dos?" he asked. And stacked the small on top of the large. I held the small to his chest, and he put his hand on my chest.

"Aquí? Or aquí?" he asked, switching his hand from left to right on my chest. I held it against the left side, and put my hand on his chest, left side, and he said something. I'm not sure what, but what I sensed was it meant we were heart to heart, connected.

His eyes and his small hands explored me then, all my details — little fingers tracing my arm tattoos, my scars, boldly feeling into the deep groove of a scar from where my right hand had been re-attached. He also touched the birthmark on my forehead — a red blotch that seems more like a marking than an accident of phenotypic development. All these marks, all of which make me distinctly me, he asked about with genuine curiosity. With his mom as translator, I answered his questions.

I stood and he reached up to me. I held his featherlight body in my arms and he lay his head on my shoulder, nuzzling my neck. He touched my pendant several times before I said, "You want it?" His eyes sparkled in answer. And just like that, this most treasured possession, I gave to him. On him, the pendant looked larger, and he grasped it and held it up for all to see. "¡El corazón!" he shouted. And then I knew. To heal my flesh and blood heart, I had to give my heart away. I had to give my heart to heal it. Of course.

Here was love. Raw and pure between us. Not like any love I had known before. Not romantic. Not parental. Not even familial.

It was instead elemental presence. Love with a capital *L*. Both the embodiment and the enactment of the form. The very engine of creation, and of the wild and precariously balanced universe, from which this particular expression — he and I — took its substance.

There was our shared phenomenon of finding rocks of a certain shape, and our chance proximity on a large, densely populated planet. But that alone wasn't bonding us. Rather, it was the mysterious current underneath such kismet, like rushing water but weightier, and even more powerful.

After this encounter, I told you about it, and sent you a video of him and me snuggling and smiling, my heart pendant around his neck. You responded with your characteristic enthusiasm, "Beautiful, Elizabeth, just beautiful." It wasn't an easy experience to articulate, and I did a poor job of it via text messages, but you got it. You knew. It was finding that pure, automatic, perfect human connection randomly out in the world that healed me all the rest of the way. I could feel it even then. Right?

But it's an idiom for a valid reason: old habits do die hard. My previous relationship, which began inside my failed experiment of polyamory and ended in deception and disillusionment, still pulled at me. Perhaps because of karma, or lady justice — that statuesque blindfolded woman holding imbalanced scales — the heartbreak and betrayal I helped cause for that disembodied voice was visited upon me and multiplied.

And yet.

The chemistry I had with her, the way we knew one another's bodies and found such pleasure in them — it was compelling, and I went back to it again and again. I had been seeing her at a rate of as much as once per week. We met in the middle at a hotel. I went to her apartment in Orange County. We had a tacit agreement to be friends with benefits, and I was actively dating other, new women I was meeting on the internet at the same time. You had a front row seat to this drama and were thoroughly entertained. Knowing what I wanted, and knowing it wasn't a friend with benefits (at least, not ultimately), you gently and lovingly encouraged me not to keep seeing her. And, after Dimas, after bumping up against a different kind of love, I wanted to discontinue those encounters with her. But I went back again, and again one final time before it concluded.

Then there was you, and the erotic energy that makes up some measure of our connection, and the memories of you as one of the best lovers I'd ever had in my life. Could it be we reconnected for a purpose? And was that purpose to spend the rest of our lives together as committed, romantic partners? But you were still with the same man you had been with that night, all those years earlier. That disembodied voice I watched you lie to through the phone while naked in my bed. You had other affairs, too, and at one point, you even left him. But you went back to him, and then had another affair. And that one ended with exposure. And when he learned about that affair, you told him the truth about all of them, except for one. Me. And here — a year after that big epiphany, which your relationship with him survived and grew even stronger for — we were again.

Just when I started thinking maybe you should leave him and be with me, I met my would-be next and most recent love. And the connection I found with her was a force I couldn't resist. Consciously, I knew I had a choice to make. In the end, I chose to move in the direction of availability. A direction that led away from you. And initially, my relationship with Would-Be was so rich with intellectual stimulation, creative inspiration, sexual chemistry, and adventure — it seemed I was being rewarded for my correct choice and for my healing. Because as then, where I found you now is in the same unavailable position. Unavailable for a committed and out romantic relationship with me, despite the energy that dances between us. Available, perhaps, for an affair again. It seemed the universe was testing me with these forked roads, to see which I would choose. The new and beautiful, seemingly available woman; the avoidant ex; the big love of the past, which also happened to be an affair. Perhaps it wasn't so much a test as a gauge of how much I had healed.

And oh, how I thought I'd healed.

Yet, there has been healing. The kind that happens automatically, with time and proper conditions. My participation is not required, and my scars attest to the fact I am healed. And even though, three months into our relationship, I discovered Would-Be was lying to me and cheating on me with a woman she met on the internet — what if this doesn't have to mean anything? At least not anything beyond what it means, which is that Would-Be falls in and out of love frequently and easily. What if the fact remains I have these scars, and they stand in bold, dark testimony?

Kintsugi is the Japanese art of repairing broken ceramics by filling the cracks with a gold-dusted lacquer, leaving a glowing golden seam that highlights rather than conceals the wounds. This reflects the philosophy that our injuries — what we've suffered and survived and learned from — make us richer and more beautiful than we were when our skin (and hearts) were perfectly smooth. Experience makes art of us all. We are not destroyed by damage, instead we are refined by it. And like a kintsugi pot, asymmetrically and perfectly threaded with gold, I am whole.

LITTLE DEATHS ALL IN A ROW

Death, and After

What do you think happens after we die? It's my favorite question to ask strangers I want to get to know, whether in a friendly way or a romantic way. It's an unconventional icebreaker and people are often startled by it. Once, a woman was offended, and immediately didn't like me. A few times, I got one of two of my least favorite answers. One: we go to heaven. Two: nothing. The reason why these are my two least favorite answers is because I know in my marrow neither is true.

I have always been afraid of death. The few times I interrogated that fear closely, I came up with nothing beyond the obvious: loss, the unknown. When my grandmother was dying at the age of ninety, her eyes were bright with life and sky-blue. I held her hand and looked into her eyes and I asked her, "Are you afraid?" Without hesitating, she answered, "No. I'm ready." We talked more, something about how much the world had changed since her childhood, how her father had to walk two miles in each direction barefoot on a dirt road to get a bag of flour. How she made only ten dollars per week at her first full time job working at the Number 2 Pencil Factory. But what I couldn't stop thinking about, and couldn't quite believe, was her bold assertion. Her fearlessness.

///

Weak hands curl and stretch at the end of chubby new arms when we come into the world. Weak hands tremble at the end of withered worn arms when we are about to pass back out again. On these edges of life, grasping isn't possible. Not yet. Not anymore.

The cruelest aspect of our humanity is also the loveliest: duality. Left hand, right hand. Feminine, masculine. Light, dark. How electric and soft is the touch of fingers tracing down our backs then back up again. Which role to perform, male or female? They expect that we perform the role assigned to us at birth. The role they have to peer between our legs to determine. All babies are undifferentiated in every way but for this small part between the legs. Elders become undifferentiated in similar ways if they live long enough, hosts of death. Where the dualistic nature of our three dimensions blurs is where grasping cannot yet or can no longer occur.

Three dimensions are required for grasping. Axis x, y, and z. My hands stretch out to the sides, horizontal. My hands stretch up to the sky, vertical. My hands have a depth filled with bones, blood, muscles, and vascular structures.

///

She died on my night. When I was younger, about thirteen years old, my mom had a client through the Hearing and Speech Center nonprofit she worked for who was deaf and blind and lived alone with her dog, Skippy. Her name was Kate. She had no family. Her husband had left for work one day about ten years prior and never returned. Her son was in prison for armed robbery. My mom adopted her into our family, and she became like a third grandmother to me. She loved to tell me stories in her deep, raspy smoking voice. A Virginia Slim floated perpetually before her face and a veil of smoke always wrapped her head. She died of lung cancer, which had progressed so far she had an actual hole in one of her lungs. There was nothing medical science could do to help her, so they recommended hospice. She came to our house to die. She slept on the kitchen couch (yes, we had a couch in the kitchen), and my mom and I took turns sleeping beside her on the floor each night.

It was around 2 a.m. when I woke, inexplicably, and sat up. I looked at her, and saw she was sleeping peacefully, silent but for the ragged sound of her inhale and exhale.

Space began to open up between each breath. The silences grew, interrupted only by her less frequent breaths. I counted between each breath. Ten seconds. Twelve. Seventeen. When I got to thirty-four, I realized another breath would not come. I had watched her take her last. I had been looking at her chest, waiting for its rise and fall. When the next breath didn't come, I looked at her face and it was not her face. It was completely changed. The importance of that fact was to teach me the first thing I would know about life and death: Whatever we are, we are not these bodies. What we are inhabits them for a time, then leaves them like a crab leaves its shell. The body, uninhabited by life, decays and turns to dirt. What it is that leaves and where it goes, and whether it contains what we know as consciousness, and whether it remains individuated — these are my smoldering curiosities.

///

Dexterity of movement when fingers bend, turning hands into shapes that represent words and phrases. I grew up with American Sign Language (ASL) all around me. My maternal grandparents were Deaf, and my mom's first language was ASL. She worked as an interpreter. I watched her work: on television in the small square at the lower right of the screen, on the side of a stage, standing beside a speaker at a podium, seated in a chair. Usually, she wore a black top to sharpen the contrast of her pale hands. This made her hands seem to glow. Her face, too, seemed filled with light as she moved and contorted it with exaggerated expressions that serve as the grammar of the language. She was mesmerizing: Her hands flew, fingers creating rapidly changing shapes, carving language into the air, throwing dancing shadows against the wall that looked like tiny wild animals, a spatial menagerie.

///

When people claim they're not afraid of death, I don't believe them. I suspect, instead, they simply haven't thought much about it. What I counter with is, Are you afraid of your mom dying? Or your child dying? Or your spouse? Every time, the answer is yes. Case closed. Because when you die, it's as though everyone you love dies at once, so how could you not be afraid of that?

When I learned more about biological systems and how they're composed of cells that divide and eventually decay, there was one consequence of this fact that seemed to prove the survival of consciousness beyond death. If a body is a whole new body every seven years, where 100% of the cells are distinct from what they were seven years prior, and consciousness gets conveyed across and between those distinct bodies, then consciousness can be conveyed from one body to another. And yes, I'm aware of the gaping hole in this theory, which is that after programmed cell death, there is no copy being made of the last cell. There is merely death. If it's true we amount to no more than a mass of cells held together in meaningful patterns that somehow make these brilliant organic entities with lungs that breathe and hearts that beat and eyes that see, then once those cells are gone, they're gone. Consciousness can be replicated across bodies while alive because the new body is an exact replica of the old. Memories can be carried along in the replication.

This may be why American culture largely ignores death, denying its existence, so that we don't have to walk around feeling our mortality in every exhale. Feeling the mortality of those we love in every embrace of their animated flesh. Published sociological research supports the century-old theory there is a pervasive death-denying culture alive in the United States, yet there seems to be an unnoticed value to holding the fact of our mortality in mind: Time is finite, and therefore precious. One short life is nothing against the backdrop of history, that of humanity as well as of the planet, whose origins we can only guess at. Were our lives infinite, not only would that lead to a disastrous overpopulation problem, but imagine the apathy we would feel. Already, with my mom having turned seventy-seven, I don't call her every day and visit her every month because I take her presence in this world for granted. I believe we have time. Only occasionally does it occur to me with alarm her time here in this realm — with her voice and her grey eyes and her hands — is terribly short. And so I book a ticket for me to go there or for her to come to me. If she and I both were going to live forever, I might never speak to her or see her.

///

When I was small, under the age of ten, and my hands were about half the size of hers, we made the sign for *I love you* with our two hands stacked. Mine palm side up on hers, middle and ring fingers folded in, hers beneath mine, same fingers folded over. We had our hands photographed this way, captured in black and white. Around the edges, a gathering darkness makes our bright hands look triumphant. Like we'd overcome something together, or like we'd won, and this was our victory grasp.

Now, a few decades later, I can make the same photograph with my children. Preemptively claim a similar victory with them — our love-shaped hands chasing back the darkness. Perhaps my mom did it with her parents when she was small, in a time before photographs were easy to take. Her father would have wrapped her small hand in that way, and they would have smiled. And my mom would have thought them triumphant, thought they had won, even well before the game was played out, but the end would come too soon and with defeat. A few decades later, when I was growing in her womb, in her fragile bowl, her father died by suicide. He gripped an electric circular saw and pushed it to his neck. The trigger of the saw is similar to that of a gun — you press with your index finger while gripping the handle to make it work. And when the saw made contact with his neck, his dexterity would have left him finally. His hand would have opened, dropping the saw. He used his own hands to take his own life. He gave his body over to his hands.

But perhaps that doesn't equate to defeat. Perhaps the gathering darkness around our love and light-filled hands is as friendly and unmenacing as its counterpart. What if both the dark and the light are brimming with gifts?

Sometimes, I look at the photograph of our glowing hands filled with light and I feel it inside me still. Like hope.

/ / /

All good things, and bad, must come to an end. The tyranny of unrelenting change is that it pummels me with loss and simultaneously sweeps all hardship away. It's the ultimate paradox in a world filled with paradoxes: Time is the biggest curse and the biggest gift. Time is cannibalistic. Time is the great neutralizer of all pain. The question of whether some immortal ghost version of me that exists before and after me, that can travel outside of me, and that watches my life play out through my eyes — that has me on the edge of my seat. I want the answer to be yes, and I have had so many experiences in my life that seem to suggest this as a reality.

And yet.

Their cheeks were too red. The genderless worker in the movie theater in an old-timey uniform behind the concession stand in the lobby. I walked up and saw, displayed all in a row on wall-mounted shelves behind their head, objects whose wetness and redness initially perplexed me. But then I knew, and of course! Hunks of raw meat took the place of popcorn and candy.

"I'll have that one," I said, pointing. The genderless worker pulled wax paper from a roll and grabbed my hunk of meat, half-wrapping it, then handing it to me. A bit of its juice spilled over the edges of the paper and onto my hand. As I walked into the theater to take my seat, I inspected my meat more closely. Maybe it wasn't completely raw, but it was cooked extra rare. I sat down among others in their seats eating their choice cuts. As I waited for the movie to start, I felt a tickle of anticipation, and I took my first bite. I would sit there and watch the movie while consuming my hunk of meat.

This was the hallucination — the waking dream I had while meditating for three days, eight to ten hours each day, with my arms held up in varying positions for durations of an hour at a time. What had me there doing such nonsense that put me in touch, at least metaphorically, with a world beyond death was death's counterbalance: sex. Not directly, but I was there with my lover at the time, my tenth love, doing what she wanted to do, what she believed would help us in our relationship. This kind of meditation, she said, created the equivalent healing as thousands of hours of talk therapy. And what we probably needed instead was all thousands of those hours.

I've cited this waking dream multiple times to friends with existential anxiety, and each time, I'm endlessly delighted by the symbolism. Our bodies are pieces of meat. This became clear when mine was ripped apart during a motorcycle accident that happened like all such things happen: out of the clear blue. Surprised as I was to find I'm just as subject to the laws of physics as anyone else, the most alarming feature of finding myself there on the hood of that car in terrible pain was the sight of my own right hand hanging off by skin, with splintered bone and blood protruding where it used to be. What this made me painfully aware of is humans are animals. We are pieces of meat. And whatever it is we really are — this soul this ghost this consciousness this witness — it's literally consuming that piece of meat. It's using it to take a ride on this planet and to experience itself as individual and relative. We can only know ourselves in relation to other people, places, and things. We can only know anything by its relative nature. We can know pleasure only in contrast to pain. Good in contrast to bad. Love in contrast to hate. Young in contrast to old. Hard in contrast to easy.

///

After my hand was torn off, vascular surgeons worked meticulously for hours overnight piecing back together each tendon, ligament, vein. I picture them sometimes, backs bent, magnifying glass over one eye, bright light illuminating the violent red under my skin. I can see it so clearly, it's almost as if I'm there. As if they could pause, look up at me, and I could see their faces. Their careful work was purposeful. They worked with the intention of restoring the feeling, functionality, and dexterity to my right hand. Later, with a cast over my arm and my fingers poking out from the top, I worked with an occupational therapist named Tina. She massaged my numb fingers and thumb for a while then gave me tasks to perform with my injured hand. Place pegs in holes, squeeze things, hold various objects. The hardest task was to touch each finger in succession to the tip of my thumb.

My thumb felt dead. No feeling, no ability to move it. If I tried and strained very hard, I might see a twitch. Otherwise, it was unresponsive and insensitive. I moved each finger as far as I could, and I couldn't reach even my pointer to touch my thumb. Day after day, I tried, and one day, it suddenly happened. I could touch the first couple fingers to my thumb. Then all four. The day I first touched all four fingers to my thumb was the first time I remember feeling deep, round, palpable hope. I'd been afraid I wouldn't be able to use my hand ever again. But this gave me hope I would have full use of my hand again someday, even if not full feeling.

/ / /

During another of those physically challenging, arms-up meditations, I saw what it is that we really are, and what happens when we die. In the last five minutes of the sixty-two-minute meditation, my body felt on fire with pain. Every muscle was screaming. That's when I saw it.

My tenth love turned translucent and inside her I could see tiny lights like flashlights at each chakra. From each emanated fluid-like ripples that came together and expanded out. When we die, the lights collapse into one and eject from the body and it feels like coming out of a sixty-two-minute arms-up meditation and lying down. Lightness and relief and ecstasy and bliss.

It's tremendously liberating because, yes, it's hard to be in a body. They're dense and spongy and heavy and painful. They absorb everything around them, which is why it's so important to move and stretch, to literally wring the body out every day. And sleep is very important — it's the time where this fluid-like light being we really are can recharge by going back to wherever we came from and gather the strength to be in this body another day. The forces at play on our bodies every moment are so much pressure. To begin to fathom this, imagine the earth spinning as it hurtles through space around the sun. The surface of the planet is moving at approximately 1,000 miles per hour. And underneath, dynamic play between the layers, from the deepest interior to the surface, exchanging angular momentum. That spin, counter-spin, velocity, combined with the orbital thrust and the whole solar system moving as a unit — all that speed and motion and force — it's a lot.

Releasing from that arms-up meditation was like the light-fluid releasing from the body. It's what I came close to experiencing on the hood of that car. But it's not the whole story. At first, I thought what I was shown gave me the gift I've been waiting for: to understand death. But since that day, I've had vivid dreams every night and now I know there's so much more. What I was shown scratched the surface, yes. But even the surface is deep. And all the rest is beyond fathom.

///

The walls and the floors and the ceilings of that room were beige or gray or white — something unnoticeable. Something not memorable. The tray table was always near, always covered with items. Supplements, a book, water. Mundane objects there for mundane hours and days stretching into weeks. The sounds and the smells and the television affixed to a big arm affixed to the bed that we hung a makeshift sling from to hold my bad arm, the one with the re-attached hand, which hurt the most. Electrical storms of nerve pain exploded up and down the arm in throbs. Pain was my rhythm, my music, my companion, my hated friend. The only friend who never left. The television that held the sling for my arm was also beige or gray and not memorable and there were flowers here and there and other bursts of color in pictures my kids drew for me and in crystals people gave me and greeting cards wishing me a speedy recovery. A numb mind and numb hand together made for a numb existence punctuated by podcast-listening and pain and novel-reading and phone conversations and pain and hand therapy sessions where I touched my fingers to my thumb and felt hope.

But hope for what? Hope to regain the ability to grasp? And why, as human animals, do we have the urge to grasp and try to hold on to bodies falling through space while we, ourselves, fall through space? Slipping through space and time as a fish slips through water, its muscled body undulating its passage through depths unknowable to us, land bound as we are.

///

It was a Tuesday evening just after dusk. I was on my way to meet a friend before a recovery meeting I regularly attended. I had been talking with Michael, one of my co-parents, and had trouble ending the conversation in time to leave when I'd planned. (How intricately I would replay that conversation after, wishing it had been shorter or longer, just enough time in either direction to avoid that one millisecond of inevitable collision.)

I hopped on my still new, yellow Ducati and rolled, savoring the feeling that always filled me when I straddled that beautiful machine. The joy in my bones, the lightness, the freedom, the sense of oneness with the motorcycle — an exquisite instrument, an extension of me with agility and articulation and acceleration and speed. I would say it was like flying but it was better than even that because there's propulsion and angles and torque and buoyancy: an entire physics facilitated by the machine, by gravity, by the road — by the interplay of all these — that can't be duplicated.

I was dressed in black lace-up boots, jeans, a t-shirt, a hoodie, a leather jacket, leather gloves, and a full-face helmet that matched my yellow bike. I was tense and loose at the same time when I saw a large, white SUV casually pull out from behind a stop sign on a side street. I quickly calculated that it would be directly in my path in one second.

My body stiffened and heated; sweat immediately surfaced on my skin. I closed my eyes, and then impact, impact, impact, impact, stillness.

On the hood of the car, I felt myself bleeding to death. A man saved my life. He slowed my bleeding enough to help me survive until the ambulance arrived and the paramedics gave me a blood transfusion, the first of many. Other people gave blood from their veins so the paramedics and then the hospital could put it into mine and help me survive what should have been a fatal accident. My son was three at the time and my daughter was eight. My biggest fear around death had to do with leaving them and missing the chance to see them grow. I was also afraid for them to miss the chance to talk to me and interact with me and feel my arms around them and hear my voice speaking to them. I was afraid of the loss of continuity and stability that would have resulted from my abrupt disappearance.

These acute fears in that moment caused me to panic. While I was living through a moment that should have been my death but for the grace and luck of intervening helpers, my response was to scream and hyperventilate and panic. And perhaps it doesn't matter what state of mind one is in when one dies, but what if it does matter? What if the quality of one's death matters and affects what comes immediately after? And because I don't want to be panicking the next time I meet death, but instead have some measure of peace in my mind, I searched for a way to overcome my terrible fear of death. It would be the same way I overcome anything I'm afraid of, which is to get very close to it and intimate with it. And so I volunteered for hospice. I signed up as an energy healer, offering to give Reiki to people while they are transitioning from life to death.

///

We are bound to land. Bound by gravity and other forces. Forces of attraction and repulsion. We obey the laws of physics, chemistry, and biology without thought or consideration, just as our autonomous bodies, our vastly intelligent and intelligently designed biological systems beat and flow and self-correct and repair and divide and grow and heal and die off and shed without our thoughts or participation. Who then are "we"? What is this "our" I speak of, if not the body? If not the biological system, which includes the brain and its electrical impulses creating these thoughts and connecting them in meaningful or meaningless ways, depending on subjective opinion. Which includes the hands, designed to grasp.

///

Part of what I learned by showing up for people who are facing death is that they, too, fear it, and so do their loved ones who will be left behind. This surprised me. Somehow, I thought being right up against it would require a kind of reckoning that would inevitably make fear yield to peaceful acceptance. The very opposite is more common. Speculation on what happens, hopes, fears — they all come bubbling to the surface from deep below where they've stayed long buried, and stir up more fear. And I will tell about my hospice clients and those individual, very distinct journeys. But there was another place I was drawn toward in my effort to befriend death: the metaphysical arts. People who claim to be gifted with the ability to communicate with those who are deceased. I sought them out, purchased readings with some, and with others, simply sat them down and asked them questions.

My sense of wonder and woo is perfectly counterbalanced by scientific rigor and skepticism. Even with those hallucinations I had while meditating and how real and powerful and correct they seemed to be in the moment, it's never long before I doubt the credibility of such information. More likely, I decide, it was some vivid daydream induced by chemicals produced in my body. Not unlike the effect of a drug, which, when under the influence, one's witness testimony can never be counted. What I also understand as a scientist involved in clinical research and drug development, is there is vastly more unknown than known about the universe by the human mind. Even our own biological systems evade our complete understanding. In fact, most of our effective Western medicines are products of trial and error. While testing for an effect on one indication, such as heart disease, a separate yet unexpected effect is observed on a completely distinct indication, such as anxiety or depression. This patient still has undiminished heart disease, but she's happier and feels a sense of calm she hasn't experienced before. Therefore, let's seek commercial approval for *that* indication. More often than not, a drug's specific mechanisms of action are not completely understood. And, neither is the mechanism of disease. What Western medicine has discovered is the body's own immune system does a much better job at preventing and treating disease than what our best interventions have achieved, save the two most important inventions in medical history: antibiotics and vaccines. Important in that they "save" lives that were, before the discovery of these treatments, unsavable.

As a person who was pulled directly from the jaws of death by modern medicine, medical technology, and surgical arts, I do appreciate these discoveries. And yet. What I know is our bodies know more about this phenomenon of healing than we do. And we do our best intervening when we leverage that intelligence. How to save a person who has lost a fatal amount of blood? Infuse more blood into their veins. Blood taken from the bodies of other people. Not anything we, ourselves, could ever dream to make. How to treat incurable cancer? Leverage the body's own immune cells and enhance or proliferate or modify them in a way that targets some key signal in the mechanism of disease, like a steel rod jammed into the spokes of the proverbial wheel. This fact, that the best scientific minds have yet to come up with anything a fraction as brilliant as our own blood, our own immune systems, our own symphony of intricate, autonomous organs working in perfect rhythm to self-correct, to heal, to grow. And yes, eventually, to stop.

///

Perhaps there is a wisdom in seeing its design as purposeful. Perhaps labeling "grasping" as negative or as an undesirable action is faulty. When I was a new human, smaller and made up of 100% distinct cells (because they die and are reborn, making the body an ever-changing entity), I grasped my mother's finger, her breast, her pant leg. I grasped objects and put them into my mouth. Hands and mouths are for exploration and discovery. All new humans instinctually know this. It is this primal knowing we lose along the way. We are taught, as we grow, what to question, what not to, how to behave, what is right, and what is not done. These rules of socialization replace the primal knowing and we stop grasping. We stop using our hands and mouths for exploration and discovery. And although I no longer want to discover more about a centipede by grasping it with my hand and placing it in my mouth as I did when I was a new human, I want to discover more about her that way — the woman who has captured my attention and drawn me into her mind. Or more, her metaphysical heart. Perhaps her soul. This force of attraction I feel makes me want to use my hands and my mouth to explore and discover, uncover what more is there. What sensations. What experiences. Depths perhaps previously unknown to me. Perhaps I was designed to grasp, and I should trust my instincts and my primal knowing more than I trust my cautious, mannered thinking. Perhaps I am meant to grasp and let go, grasp and let go, grasp and let go. This closing and opening of the hand is the natural rhythm of things, of a body falling through space alongside other bodies falling through space.

But what about those sea creatures? All the dark and bright water. The depths to the shallows to the unknowable deep. Its creatures arrive and depart without hands. Slipping through time and space —

creatures like scallops, whelks, oysters, clams, mollusks — they never grasp.

///

On my right arm, on the underside of my forearm, there is a neat seam from wrist to elbow pit wherein titanium rods were fitted to replace bones and connect my severed hand back to the soft tissues keeping it alive. The skin of this re-attached hand is thinner and more fragile, easily cut or rubbed off. Perhaps it's a compromised blood supply, or merely the life drained from the skin when it was separated from its source. The shape of the hand, too, is different. My palm is narrower, the muscles in it less defined. Nevertheless, it functions. It senses. A hand resurrected from death, total numbness, complete paralysis. Facilitated by the interventions of the best medical technology, talents, and minds, my body worked its magic and repaired. Skin grew back together and closed. Nerves regenerated or forged new pathways. With the conditions optimized, my body did what nearly all bodies do unprompted: healed. The seam on that underside of my forearm has always reminded me of a power line, subtly curved as though from gravity, as though with the slight weight of birds. Birds perched. At rest yet alert and ready. Their whole countenance vibrating with the imminence of liftoff, of flight. Skyward, up, up, and gone.

THE HOUSE
OF
DESIRE

A Penis Just Works Better

Did I ever tell you, my tenth love, about the time I used my hip bones as weapons? It was about my early experiences with sex.

Sex didn't involve much pleasure for me at first, even after it was real, consensual sex and not just violence. Every time I was drunk, which was often, I wanted to try it again, certain I was missing something; some key ingredient that had people fussing over it so much. When I did remember it or parts of it — which was rare — pleasure never factored in. Pain often did. And sometimes, shame. I always remembered the gross parts, like washing semen off my hands, thighs, stomach, or spitting the brine from my mouth, then rinsing with strong mouthwash. Perhaps my experience with boys and men was consistently negative because, even then, I preferred females. Or perhaps it was that male stench of entitlement and smugness.

You had a different experience and you told me about it, more than once. You enjoyed your first sex. It was with your long-time boyfriend, even though you were only twelve or thirteen at the time. His penis was small, and you barely felt anything, and it was over so fast, but you both enjoyed it and wanted to do it again. You did it in his bed, under some covers.

These are the earliest pieces of evidence I would like to submit for the record, in an effort to understand who murdered our connection, and when and how. Its death can only be a crime, such that it stands out among the epic tragedies and injustices of my life.

When you told your childhood best friend you were in love with a woman, she couldn't believe it. "But you love penis," she'd said. At least, that's what you told me she'd said. Then, there was the dinner you took me to at a French restaurant for my fortieth birthday. I had been talking academically about a previous love and about sex.

"There was something maybe about the forbidden nature of it that made me feel almost addicted. I had more orgasms in a row that day than I ever had before or since." And it must have made you jealous because you took the same detached academic tone when you told me with curiosity in your eyes, "Yeah, you know, in terms of the number of orgasms and how long it takes to get there, a penis just works better for me." Even after knowing you for six years, I hadn't yet realized how vengeful you could be.

When we were walking the streets of Paris in 2017, we came across a women's march against sexual harassment and assault. You commented on how too many women were making something out of nothing, and that ass grabbing didn't mean anything. Later, in our hotel room after we finished having sex and before we fell asleep, in the dark with your back to me, you told me with some shame in your voice about your secret fantasy, which was to be raped by multiple men at once, erect penises poised at every orifice. I wondered if only a woman who had never been raped could harbor such a fantasy.

But before all of this, our life as lovers on this planet began. Because I had known you so well and loved you unconditionally, and because of the intense sexual chemistry we discovered living between us (Had it evolved or was it there all along, adding to the complex richness of our connection we chose to call friendship?), these red flag moments that would have sent me running from another lover washed over me almost unnoticed.

It is theorized life began on this planet with single-cell organisms that reproduced asexually. More than a billion years of evolution later — so a fossil record with gaping holes shows us — sexually reproducing complex organisms showed up on the planet. This jump was potentially facilitated by volcanoes and unearthed amino acids but also, and of course, magic and luck. Along with the phenomenon of sexual reproduction came aging and programmed cell death. Life produced from sex must necessarily be finite.

Perhaps that's why you said it when I invited you to be my lover: "Romantic relationships end." Relationships involving sex must necessarily be finite. Later, we joked about how I had prepared an objection handler and was ready to handle any objection save one. If the feeling wasn't mutual, that would be a full stop. "It's mutual," you said then, holding up your hand to show me how it trembled. "And I'm terrified."

The sex we had in those first eighteen months was some of the most enjoyable of my life. So easy, so full of love, so hot. And we never stopped having sex, but we did somewhat settle into a routine as we approached year two, wherein familiarity with one another's bodies unconsciously scripted exactly how to produce the desired outcome, often the most efficaciously, which led to a sometimes boring going-through-the-motions performance. In flashes though, always, my resistance to intimacy, or yours, would crack; we would know ourselves as we truly were and reach ecstasy again, and again, and again. That surrender, that light, that beauty: it's how I choose to remember you here, now.

Allow me to find a lens through which I can relate to you. One that will allow me to see a time when my internalized misogyny made me eroticize violence. A time when a penis just worked better for me, too. To tell you the story though, I must start with an allegory about a bowl.

I have a mended pelvis: a very large screw through my left sacral-iliac joint (so long it reaches across my back to almost touch the right SI joint), metal plates and pins like small nails replacing my crushed pubic bones and pubis joint. Nails and screws in bone as wood. The materials for carpenters to fix bodies or chairs, making them useful again. Stronger, even, reinforced with metal. Mettle.

Breaking it was easy, as breaking our connection, in the end, was easy — as breaking things tends to be. It's the putting the pieces back together and healing that takes skill and time and infinite patience.

I own a ceramic bowl that was a gift from my oldest friend. She made it with her own two artist's hands. Inside of it is a ceramic red heart. The bowl has traveled everywhere with me, every place I've lived, for nearly two decades. One day, when I wasn't home and accidentally left my bedroom door open, my dog, who was mad at me for not being home, went into my bedroom and knocked the bowl down from the top shelf of the bookshelf onto the floor. It broke. I pieced it back together as much as possible. Glued it with superglue. But there is a sliver lost from the side that I can't replace. And the little red heart, though intact, is no longer affixed to the inside of the bowl.

The pelvis is a bowl, a precious bowl I carried everywhere with me for four decades. Many times, I was reckless with it because I sat it atop motorcycles and rode them at high speeds, zipping alongside equally fast-moving concrete and metal. In the same accident that tore off my limb, when that inevitable day came and my motorcycle smashed into another vehicle, I broke my bowl. Just like that. I laid the bike down to the left and my left sacroiliac joint came apart as though it were a Lego piece that snapped off. I collided with the car and zoomed crotch-first into the steering column, crushing my pubic bones. But this story has already been told. What hasn't been told as much is the way it feels now, nearly four years later, to walk around with this metal pelvis, which carries this memory and its whole history inside it, sloshing like soup.

The invisible bowl made by the pelvic bones that holds its history like soup never used to be a thought in my mind. Only once had I ever contemplated those bones and the shape they make, and that was when I was pregnant. Filled with apprehension about passing a human body through it, I was happy to hear the bones were hinged and flexible. In my medical report, how my pelvis broke is described like this: *It unhinged and opened up like a book.* They call it an open book fracture. When I read this in my medical report, I couldn't picture it. How could my pelvis open like a book inside my body? Inside my skin? It seemed anatomically impossible. It made me think about how we must break the spine of a book to read it. The more we read it, the more broken its spine.

Did I tell you how I took a book off the shelf and a pillowcase out of the linen closet and tried to approximate the event by placing the book inside the tube of fabric and then opening it? But it didn't work. The model wasn't to scale. There was too much space around the book inside the tube of fabric. So, with a vague and aching desire to wallow in the trope of heartbreak and physical break and sad stories and open books, I took more books off the shelf and opened them all and scattered them around the floor. I'll never forget that: how I looked down at my open-book-strewn bedroom floor while my pelvis ached.

When my pelvis was put back together, it was held in place by a metal external fixator: basically, a crowbar attached to my hips. The pins went through my skin, like skewers driven through meat. The bar made it impossible to wear pants, to wipe myself, to properly clean myself. When I started cleaning myself again as part of my occupational therapy, I rolled into the large, wheelchair accessible bathroom and closed the door. I had a pile of clean washcloths, and I would wet one with water and one with water and soap. I used the soapy one to scrub my face, under my arms, my chest and stomach, and wherever else I could reach. But both my arms were broken and casted, and I couldn't reach my genitals. Most of the time, my genitals were cleaned by healthcare professionals, but in the rehab hospital, they didn't do a good job of this. So once, I asked you to do it. My romantic partner. In addition to cleaning out my sore, still very bruised vulva, you actually used your fingers to pull poop from my butt when I was constipated.

You can't imagine how much I didn't want my romantic partner involved in these tasks. But because you did, I felt so deeply loved that I felt even more afraid of your leaving me. How could anyone else ever make me feel that loved? And because you did these tasks, I came to believe you never would leave, because anyone who loved me that much would certainly never leave me. But alas, you did leave.

My metaphysical heart broke when you left. Like my pelvis. Like my ceramic bowl. So many easily broken things fill me and surround me. So many pains must be taken to fix them. And even in their being fixed, they will never be exactly the same as they were when they were intact. Just before they were easily broken. And they will forever hold their memories of breaking, their histories including everything that came before the breaking, including the early trespasses, the unwelcome touches, the chipping away or the wearing away, or the not being noticed or appreciated, or just simply taken for granted.

I know I promised to tell you the story about the time when a penis just worked better for me. But first, I have to tell you about the first time I almost died.

Weight is a function of gravity. The denser the matter, the bigger the gravitational pull on the matter, the heavier the matter. Our bodies are matter. One of my favorite places to put my matter body is in water.

Denser than air, water lessens the gravitational forces pulling down on me. When I float in water, I feel held and cradled like a small child. Like a small piece of matter.

Before I learned to swim, when I was three or four years old, I was playing in a hotel pool with my brother. Our parents were poolside reading newspapers. We were taking one of our family weekend trips to a hotel, which my parents would do when my older three siblings went with their father for the weekend, which was infrequent. In the pool, I wrapped my arms and torso around a blow-up ball and floated on it. Kicking my feet, I drifted over to the deep side of the pool, bobbing like an orbed fruit at its surface. Somewhere in the deep end, the ball slipped from under me and shot away. I fell into the water. I panicked and tried to scream. I thrashed around and inhaled water. Once or twice, I broke the surface and screamed for my brother, who was in the shallow end. I glimpsed him there, frozen with fear in his eyes. My parents hovered at the pool's edge, miming swimming motions, encouraging me to try to swim. I fell back under, inhaled water again, and felt sure I was dying.

And then, just like that, I was lifted out of the water in the arms of a stranger. After I finished coughing the water from my lungs, I looked at him. He was fully clothed in a button-down shirt and dripping wet. His glasses floated on the surface of the water. A pen was clipped to his soaked breast pocket, pasted to his chest. He held me on his hip and asked me if I was OK. I nodded and lowered my head, feeling shy in his arms.

You must wonder here — Why hadn't one of my parents saved me? Some years later, I quizzed my mom about this mystery, but not my dad. Because he had been, by and large with a few notable exceptions, self-absorbed and neglectful, I somehow didn't question his decision not to save his youngest child from death by water. My mom, though — her same decision sincerely perplexed me. "I was covered in poison ivy," she offered in her defense, "and my arms were bandaged and I was under doctor's orders not to get wet."

Perhaps it goes without saying that I found this defense grossly inadequate. As a mother myself now — poison ivy or third-degree burn, bandages or no, doctor's orders be fucked — I would never hesitate to jump in and intervene with the extremely preventable death of either one of my children. A mere couple of feet from me. Right before my eyes.

"Besides," she went on, "the lifeguard was there, nearby. He came running and asked me, 'Is this real?' and I screamed, 'Yes!' and so he jumped in and saved you, which was his job." My hurt and anger notwithstanding, I couldn't bring myself to remind her that lifeguards typically wear bathing suits. Not business suits and eyeglasses.

Alongside that, you should know, my mom has parented and loved me well. I am alive today in part because of that fact. She is my original, deepest, most real human connection on this earth, and she is a heart-of-gold, selfless kind of person. She did not have an easy life, nor did she have fulfilling romantic relationships with her two husbands. She worked hard and raised five children and I never doubted her love for me. Her attention was spread thin sometimes for good reason, and I forgive her everything and am eternally grateful for being made by her.

After that, I learned to swim. I learned to save my own life. By the age of seven, I joined my first swim team. For ten years, I swam competitively and in high school, beat an NCAA championship record for the one-thousand-meter freestyle, which earned me a full-ride scholarship to a college in Florida, which I didn't accept. My feet belonged on the ground by then where I could drink alcoholically and not have to curb it to accommodate a rigorous athletic program. I went to a local state college instead because my first girlfriend was in attendance there and I wanted to be close to her.

I did at least continue to run for exercise. When I ran, the weight of my upper body was distributed through the ring structure of my pelvis, down my legs, and into the ground. Each footfall to the ground reverberated back up my legs to the ring structure. Each contact with the earth pushed its weight into me, reciprocally. This exchange of matter absorbing the weight of matter in a pleasing rhythm for a specific distance or duration became my sanctuary, my balm, my pressure valve.

When I broke my pelvis, I worried for a time I would lose that sanctuary and no longer be able to run. I tried to imagine a life without that regular release. One where I walked a lot instead, or hiked, or swam — a practice I returned to while pregnant. Knowing I would have eventually had to stop running anyway, I tried to be at peace with it. I wasn't.

Peace eluded me, and instead I lived inside my neurotic brain that leaked a steady drip of anxiety into the circuitry of my body. My imagination, a gift I historically adored because it afforded me so much pleasure, now turned against me and had me seeing vivid, limb-torn-off gore when I slept, when I crossed the street, when I drove my car. These imaginings amped up the anxiety vibrating through my body until I could no longer sit still. I had to move. Mercifully, my body healed enough to return me to my practice of running. Slowly at first, I ran three quarters of a mile one day. My metal joints strained, and my muscles were sore. Over time, I built on the meager distances until I could move my body in a slow trot across a vast two miles. After, every inch of me throbbed and hurt with the effort, but that hurt was also hope. I did it and it felt like coming home.

I promised to tell you the story of when a penis worked better for me, and this is all relevant. Because long before I broke my pelvis, and long before I gave birth, there was one other time I was acutely aware of those bowled bones.

In my early twenties, I was in the late stages of alcoholism as well as the early throes of an eating disorder: I starved myself and over-exercised. Consuming only cigarettes, alcohol, and caffeine, I ran many miles each day and completed a set of a minimum of one hundred abdominal crunches and one hundred curls with each arm holding a ten-pound dumbbell in each hand. To think of that now, half my lifetime later, I marvel at the miracle feats available to a youthful body.

I was at least twenty pounds underweight in those days and felt completely empty inside. I carried the lightness more like emptiness and then like fullness, filled with chemicals, walking down the street with helium in my bones, liable to blow away in the wind. I had long, thin, wispy hair to accentuate my waif body and I always pulled it back into a ponytail, mutinous pieces pulling free in the front around my face, the frail flyaways catching in my eyelashes.

I liked to lie on my back and gaze at the sharply protruding hip bones framing my pelvis. I thought I could use them as weapons if I had to. Slam my hips into someone and impale them against me, one blade on each side. The boyfriend I had at the time — my last boyfriend, having nearly concluded my experiment with heterosexuality — liked those sharp hip bones. He would place a large, open palm over each hip bone and get a certain expression on his face — somewhere between satisfaction and anticipation. He would sit on his knees between my legs, ready to impale me onto him, and I would close my eyes and turn my gaze inward to the hollow, chemical-filled shell I was. Always drunk or high and usually starving, I imagined I enjoyed having sex with him, as it filled me in a way that approximated buoyance, my sharp bones stabbing into him while he stabbed into me. Some kind of strangely satisfying symmetry. Once, we did it standing up, my back against the wall, my legs wrapped around his hips, and there was a fleeting moment of genuine enjoyment — of pure bliss, in fact — before it was swallowed up whole by the voracious void inside me. For several years thereafter, as I wrestled with my internalized homophobia, I pointed to that fleeting moment as evidence that maybe I could be straight, after all.

Eventually, I healed from all shreds of homophobia, and many years after having healed, I met you. Steeped in Eastern European patriarchy as you were, where misogyny was sexualized and heteronormativity included the blatant subordination of the vagina to the penis, yours was a more insidious form of internalized homophobia, and you had decidedly not healed from it. Sexual harassment was an everyday occurrence from your earliest memories onward, and you routinely asserted generalizations that demoted women and promoted men. "A man has a woman in his aura for seven minutes after having sex with her, but a woman has a man in her aura for seven years. We have to be more discerning with our sexual energy." So, in your view, only men could be sluts with impunity.

But years later, after we came together as lovers, you were the ball and I was the socket. You moved inside me and I lubricated your movement while also restricting it in favor of getting what I was searching for — greater stability. It wasn't to be found there, in that joining, and I wouldn't know until it was too late, too soft, too warm, too wet, too open to all possibilities including death and whatever comes after. I would die a little every time you moved inside me. And in dying I would feel alive in a way that could not be attained by any other means in this animal body. To die this way was to live. When I felt this, I tried to stay there, I tried to slip out through the openings in my skin to where the aliveness throbbed and disappear into it. But I never could. I slipped out of it, slipped back into my body, crashed back in like a wave against a shore, pulling sand and gravel and tumbling it around in an eternal flailing that smoothed and polished everything to an imperfect roundness, a suppleness in my hands. My hip bones, too, felt less sharp in the aftermath.

Let Blood

I don't remember why we were in New York, but it was toward the beginning, soon after we changed from close friends to lovers. My tenth love. We sat on a rock in Central Park and had a picnic. It must have been summer because I wore running shorts and a tank top. I felt the wetness between my legs, like I did a lot in those days when in proximity to you, and I wondered if it was because you perpetually turned me on or because I'd started bleeding. I sat with one leg bent and the other lying flat and rested an elbow on my knee. In a way I thought was inconspicuous, I reached around the back of my thigh and slipped a finger in under the built-in underwear of the shorts and touched the wetness. By the time I noticed you were watching me, it was too late. I'd already drawn my hand back with a spot of bright red on my finger. Your mouth bent into a smile and your eyes looked amused. Frowning, I held it out to you and said, "I'm bleeding." The unmistakable expression that passed across your face then — it surprised me. It was lust. Pure, unbridled, animal lust.

"This turns you on?" I asked, smiling, wiping the blood across the face of the rock.

"Yeah, is that weird?"

In fact, I did think it weird, but my response was no, and I felt pleased, because this seemed proof positive you were a lesbian. Maybe even a bigger lesbian than me. It might have been then when you told me the story about when you first felt sexual attraction to a woman.

///

In the late nineteenth century, it was customary for old white men to be credited with discoveries of knowledge their mothers and their mothers' mothers undoubtedly already knew. In one of these frequent occurrences, an old white man named Arnold Adolph Berthold, in the custom of so many other self-obsessed men, took an interest in testicles and what magic they held. To peek behind that particular curtain, he castrated roosters and observed the effects. Part of what he noticed in the maimed roosters was a mysterious lack of aggression. Also, their combs and wattles were smaller, floppier, and paler. "Aha!" I imagine him saying. "Testicles are what make a male animal male!"

In the next phase of his cruel and manly experiment, Berthold transplanted the castrated members of the roosters into their abdominal cavities, wherein he observed a return of the aggressive behavior as well as an erectness and increased redness in the comb and wattle. Fascinated, he killed and then dissected the transplanted animals to discover the abdominal testicles had grown a vasculature around them. This was the first recorded understanding that whatever effects were happening, they were happening through the blood.

///

In Bulgaria, during your upbringing, your family took vacations to the coast of the Black Sea where it was the custom for everyone to shed their clothes. Swimsuits were not a thing, you explained, everyone was simply naked. You joked about how one of your iconic childhood photographs was of all of you huddled nude on a beach blanket, smiling. Only there, just inside the frame and in the foreground, was an old man bent over, ass to the camera, his loose scrotum between his legs. The way you described the photo made me laugh every time, and you described it over and over, because I asked you to, and you indulged me. Usually, it would be when we were talking to a third party about some aspect of your upbringing.

"Oh, you grew up in Bulgaria? What was that like?"

"Family outings to nude beaches was one feature," I'd offer, then look at you. "Tell them the one about your family photo with the scrotum hanging in the foreground." And I'd start laughing even before you opened your mouth to tell it.

In that way, we were so compatible. We loved to talk for hours and make each other laugh. We did really well on vacation, having adventures together. Even after I foolishly told you to stop working because I was now making enough money to support us both as well as pay for the cross-country travel that allowed us to see each other. It wasn't an altogether altruistic move — I had a personal agenda, though I could argue that it was personal for us both. I wanted to free you up to travel and spend more time with me. And, for a while, you did. But after a couple of years, you considered the travel burdensome and felt controlled by me because I was the maker of the money. And I, in turn, felt the burden of what it really meant to support you financially. I had to not only provide for you and pay for your whole separate life on the other side of the country, but also, I had to support, at least in part, your son and your mom as well. On top of my own children and my whole separate life on this side of the country, it was a lot of mouths to feed and a lot of bodies in space to keep moving.

I'm getting ahead of myself though. Let's get back to the nude beach in Bulgaria when you first felt sexually attracted to a woman.

///

The story of Berthold and his obsession with testicles reminded me of my own appendage with a compromised blood supply. When my right hand was torn off and then carefully reattached, there was a long window of time during which I worried I would never get the functioning of that hand back, and the hand would be a numb, dead thing at the end of my arm, forever lost to me. And with that loss would be the loss of my sex fingers. When I first awoke in the ICU and looked at the numb, swollen, sausage fingers poking out from the wrapping on my right arm, and saw dried blood on them, my first thought was of the other context in which I had dried blood on my hand. That made me think of how I, too, had felt aroused by menstruation in just this scenario. Somehow, when a woman let me fuck her while she was bleeding, she was letting me into her innermost primal world — a deeply recessed wetland of a world with dense green jungles and vibrant, diverse life teeming with vitality and birdsong. A world where vulnerability swallowed the head of strength and then opened its mouth to reveal an entire universe where a tongue should have been.

///

You were around twelve, you thought, because you had only just had your first period. Lying on the beach, reading, you looked up to see a woman walking out of the water. Big-breasted and thick-thighed, she strode unabashedly toward you, long black hair pasted to her shoulders and chest and face, water streaming from her body. And there, between her legs, was a furry mass of black hair, tightly curled and dripping wet. But it wasn't just water on her thighs — it was blood. Blood mixed with sea water streaked down her inner thighs like the bold, long strokes of a painter. Like the long strokes of red paint in that mural we stood in front of to take selfies. You wore that red dress and hat, and I wore a white tee and a fedora. You liked me in a fedora. Now I remember what we were doing in New York, because I was braless under that white tee and that's because it was the closest I could get myself to topless while walking in the enormous Dyke March. It was Pride weekend in New York, and we were visiting to display and celebrate our pride.

When we discovered that we had left our dildo on the nightstand of our hotel room, we realized why the housekeeper had looked at us funny. We laughed. That was our first summer together, the long season of vibrant life following that of newness, of birth, of after-winter and after-death, when we first kissed. Those days were all honeymoon, all sexy, all bliss. Later, at the end, I would think longingly of those days and wonder if we could ever get back to that place. That place where we had no agenda but to enjoy each other. And where the pleasure we found in each other was mutually wanted, and given, and received. At the end, I made a music video out of all the photos from our first eighteen months, and I told you it was my favorite time. You got mad at me again, like you always did by then, because you read the very worst into all my motives, and you accused me of only wanting beginnings and newness and romanticizing that too much. But that's not what I was romanticizing. I longed for the beginning because it was when we were equals.

Long before that, though, there was Central Park and the blood on my finger and the lust in your eyes and the story about the nude beach in Bulgaria when you first experienced lust.

///

In San Miguel de Allende — an old town in the center of Mexico that looks like the lovechild of Paris and Barcelona — I purchased a ticket on a guided tour to Cañada de la Virgen. This wasn't my usual jam, but the only way to visit this protected archeological wonder was as a guest on a guided tour. The ruins, partially reconstructed, are from an ancient pyramid built by the Otomi tribe, who watched the night sky for generations and charted its patterns. They selected the site so that the moon might climb the notches of the pyramid and take the other side down, like ascending then descending stairs, in its annual cycle. Their calendar was lunar, and during solstices, when the moon aligned just so with the architecture, it was as though it sat perched atop the point.

///

You stared, mesmerized. You felt, you said, a stirring low in your abdomen and a tingling in your fingers that made you feel shy. Did you absently touch your lip, the way you sometimes do when you stare? Did your small body feel its potential? All that it would grow into? Did your future self possess you in that moment and feel shame for wanting another woman in a way you had already been taught you shouldn't?

On the last day of our first weekend as lovers, we kissed and took each other's clothes off on the floor of your living room. Shoving the coffee table aside, I turned you around. You bent over the couch, and I traced the sleek muscle ridge on your back with my tongue. All the way down to the backs of your thighs. Touching between them with the tip of one finger, I felt your wetness, and put two fingers into you. You moaned and moved your hips with my fingers. I remember the seed necklace I wore, the one my friend had given to me for protection, and how it swung in rhythm with our bodies and steadily tapped against your ass and low back. When it was over, you looked at me with a stunned expression.

"How did you do that? What even was that?"

"What?" I said, climbing onto the loveseat and stretching out naked before you. While sliding your fingers into me, you said, "It felt like more than your fingers." And I said, "That's big dick energy." You smiled that crooked, lusty smile and pushed further into me, showing me yours.

When you wanted to sell that loveseat, I had it shipped from Boston to San Diego so that I could hold onto it. To this day, when I see it in my living room, I think of that encounter, and all the countless similar others, including those when we sat facing each other and talking for hours.

///

The guide told us stories as we walked through the grounds. He spoke fast and was hard to hear, so I didn't catch everything, but the most interesting bit I did understand was the tribal leaders had a daily practice of bloodletting. Each day, one of them opened a vein and drained a certain amount of blood into the earth as an offering, a sacrifice. They believed human blood was the most magical and precious fluid on the planet, and a certain amount each day had to be fed to the earth to keep the celestial bodies moving, to keep our world turning on its axis.

It made me think of the blood I received post-accident — a pint in the ambulance then three more pints at the hospital. It reminded me of the paramedics who cut off my clothes and covered me with paper that soaked red against me, then put my broken open body onto a stretcher and lifted it from the shallow pool of my own blood on the hood of the car. The intensity of this offering struck me there, in the ambulance — how perfect strangers had gone way out of their way and undergone somewhat of an ordeal, likely with pain involved, to open a vein and let blood go from their bodies into plastic bags that then were carried aboard an ambulance to my nearly dead body, thereby, by this razor's edge, holding onto life. Thereby continuing to move in space. A body in space moving in relation to other bodies in space.

///

The big-breasted, thick-thighed, menstruating woman walked onto the sand and up the beach a stretch to a man who lay back on a large towel, naked as he came. The woman lay on top of him and kissed him, and you watched as his hands circled the small of her back. You didn't know if you hated him or wanted to be him, or both. You didn't know anything beyond the clamor of your nervous system as sexual awakening surged through its nascent passageways.

This story always made me think of my own origin story — my first perceivable sexual attraction to another female. I was around the same age you were, and I was between heats at a swim meet at a YWCA. The *W* in the place of the *M* pleased me, even then. Wandering the recesses of the building to kill time, clad only in my speedo one piece and a towel, I heard music, and moved toward it.

It's natural, it's chemical

(Let's do it)

It's logical, habitual

(Can we do it?)

It's sensual

When I arrived at the open door of the wide room from which this mesmerizing music blared, I wondered for a moment if I was dreaming, certain I'd had a dream in which I felt just this way. There, bouncing on a trampoline in a gym, were maybe a dozen older girls in leotards. They were smiling and laughing and moving suggestively, all to the soundtrack of sex.

Sex is natural, sex is fun

Sex is best when it's ONE-ON-ONE.

A pair of eyes met mine and I wanted to run, but my feet were glued in place. I had been holding the towel in front of me and when this girl looked at me, I dropped it. She smiled and, standing wide legged on the edge of the trampoline, beckoned me with the hook of a finger.

///

On social media, I met a scholar who studies ancient Mesoamerican cultures, and I mentioned this bloodletting ritual of the Otomis I'd heard tell of, in the hopes of fact-checking myself, lacking confidence as I did in both my memory and my capacity to hear on that windswept vista in the desert. In response, she sent me an excerpt from her dissertation. In it, she wrote about this very ritual and confirmed Mesoamericans did quite literally feed blood to the earth in the way a mother might nurse an infant. They believed the earth to be a living, breathing being with eyes to see and a mouth that opened and a hunger to be fed. The inside of it, they believed, was the very nexus of creation and destruction, birth and death, and this underworld could be wrathful, if scorned or ignored.

Not long ago, a friend told me about a women's retreat she'd attended called the Divine Feminine. In it, the women were taught the value and utility of their menstrual blood. Namely, that it makes both an effective face mask and a better-than-Miracle-Grow plant food. Skeptical, I asked about her experience trying out these claims.

"Without fail," she said, "the day after I do my blood mask, people tell me I have such a glow about me and that I look so good. And my queen of the night was so happy after I mixed a little of my blood into the water I gave her, she bloomed an extra time. That never happens!" She went on to tell me a mythological story they told at the Divine Feminine event where tribal women would stand bowlegged over a certain red and white mushroom as it grew from the earth and let the blood from their vaginas fall on the plant, feeding it. This ritual would make the mushroom medicinal in a way that allowed the women to benefit from eating it. The men, however, were unable to get any medicinal effect from eating the mushroom. Instead, men had to drink the urine of the women who had eaten the mushroom to get an effect secondhand. All this stirred me and moved me enough to feed my own blood to my plants, including an orchid, which was growing buds for the second time. Within a day of my blood water, six of the eleven buds were in bloom. It seemed impossible. I called in one of my co-parents (my two housemate gay men who father my children), excited to share this miracle.

"Do you see? Do you see the metaphor?"

He squinted at the buds. "I don't see any metaphors. I see some flowers?"

"Do you see how Nature is a poet and she uses metaphors to communicate with us? And they're perfect and powerful metaphors?"

He shrugged and walked out.

Short of peeing in his mouth, I realized I couldn't make him see it, this miracle. This medicine. This magic.

///

When I ran from the door of the gym with the beckoning girl flanked by the other bouncing girls, I left my towel there where I'd dropped it. In my memory, I heard them laughing as I ran, and I imagined they were laughing at me, and my pale towel where it lay, and how it had exposed me, and exposed my secret. A secret I'd been keeping even from myself — I should have been a boy. At the time, that's what sexual desire for other females meant. It didn't yet compute a girl could want a girl in this way.

Is that how you felt back then on that nude beach in Bulgaria? Did you feel as exposed, with all your skin shining in the sun?

Back to the last day of that first weekend, with you bent over the side of the couch and me right behind you — I'm thinking about it again while I walk past the living room in my house and glance at the couch, crouched there in the dark, looking nothing like how it looks in this cherished memory. Now, it looks like our shared wrath — that tall, mean place we last met. The place that looms over us now, nearly three years later. Three years ago, I told you the long distance wasn't working anymore, and that I had to ease out of supporting you financially. What you heard was, *I don't love you anymore and I'm going to abruptly collapse the floor right beneath your feet.*

It would have been worse, I guess, if I had told you the whole truth: You had become a heavy weight to carry, a great burden on me — a constant siphon of my time and money and energy with nothing to give in return. It seemed I was slowly bleeding out resources and had to keep trickling my vitality into your veins to keep us moving forward. And after you were gone, my predominant feeling was relief.

Only recently though, while going through old photos, I came across one of you, your naked lower half standing bowlegged with a smear of blood down one thigh and a bright-red dildo held in place between your legs. The brightness of the red against your pale brown skin, it shook me. It made me long for you again.

///

Prior to having my very life saved by the astonishing generosity of strangers, whose vital fluids were fed to me in my time of loss and need, I'd been largely unaware of blood banks and the practice of donors. So eager I was to pay forward the favor, as soon as I was out of the last arm cast, I rushed to my local blood bank. After filling out the paperwork and noting down I'd received several transfusions about six months prior, I was turned away, but invited back in six months more, after I'd passed a full orbit around the sun with my new, donated blood in my veins. Given blood. Kind blood. Let blood.

The last time I donated blood, I watched the woman across from me watching her phone while hooked up to a plasma machine. Plasma donations take longer, about an hour, and are mostly used for cancer treatments. They take longer because the machine separates the blood into its three main components: red cells, plasma, and platelets. I watched and thought about how far we've come since the time Arnold Adolph Berthold carved up roosters to tell the world about how there was something in the blood that maybe controls our sex drives, our moods, and even some aspects of our physical appearance. Now we know about these parts of blood: Red cells carry oxygen; platelets help clot the blood so it doesn't escape the body too rapidly or in too high quantities; plasma carries nutrients and hormones through the body.

I asked the woman working there how much blood goes to waste. The ambulance that responded to my motorcycle accident had blood on board — one more in a series of life-saving coincidences. This made me wonder: How many ambulances with blood on board don't end up needing it, and therefore disposing of it? Like a politician, the woman avoided answering my question directly, saying instead most blood donations get used within a week. She added that the shelf life of whole blood is forty-two days, and expired whole blood is treated like all other medical waste and incinerated.

It startled me to think of this vital fluid keeping me animated with myriad appetites as flammable. Then I remembered cremation. This made me think of incinerated corpses, and one incinerated corpse, in particular — Anthony Bourdain, the irreverent celebrity chef who committed suicide and was eulogized on Twitter as someone with *true big dick energy* or BDE. A viral swarm of posts about people of all genders with BDE ensued. I was amazed. I'd been using that term for years to describe myself, to jokingly explain my swagger. In fact, in the viral swarm, more women than men were described as having it, which got me thinking of this sensitive appendage nested between the legs of men and how fragile and tender they are. The irony in how this delicate part was ever equated with strength and confidence in the first place is alive, too, in the castrated rooster experiment. The force that made the wattles red and combs erect and put the confidence in the strut of the rooster was in the blood, not the testes. Instead of big dick energy, I should have been saying for years, big blood energy.

///

Before the final end, there were a few false ends — times you broke up with me then changed your mind. But, with each breaking and coming back together, the composition of our connection changed, eroded, curtailed. From the Old English "brecan," to break is to divide solid matter violently into parts or fragments; to injure; to violate. Add enough heat to a solid and the strong bonds between the molecules loosen, weaken, become fluid. Add still more heat and those bonds dissolve, too, until the matter is vaporized, free floating in the air and breathed in and out of lungs. Nothing is created or destroyed. Only found and mishandled and changed.

Three months before I dealt that fatal blow to our connection, you wrote me a six-page letter detailing the many reasons why we should break up. I didn't have the heart to argue, to attempt to break through to you. I wanted to break free of our connection that now felt like a trap. We had both broken down. I could feel it all breaking.

In response, I left, and flew home a few days sooner than planned. Two days passed before you called me in tears. You detailed three reasons why we should get back together: "I made a huge mistake! There's nobody like you! I tasted your menstrual blood!"

Blood-bound though we remain, and in spite of all the love, we couldn't survive the severe imbalance we were already experiencing. The energy can't just flow one way. It has to be reciprocal, like the platelet machine.

We broke for the last time in November. In late December, I went to Boston and stayed with a friend. During my last night in town, I invited you to come stay with me and my fully restored sex fingers in a hotel for a breakup ritual I hoped would have us end on a high note. You invited me to go do yoga with you. We had breakup yoga and breakup sex. But you were not yourself — you had not been eating and were so skinny, your eyes looked bigger in their sockets, your facial bones sharper and brighter under the skin. Your big eyes seemed to dart, panicky, ricocheting off surfaces like bullets. It looked like it should have a sound effect: *ping, ping.*

I wished to not let that last painful sex with tears and your already emaciated body be the end. I wished for an earlier sex to be the end. I knew just the one. It started in your living room, which was wrecked in the aftermath, and ended in your bed, with your fingers inside me, my back arched over your other hand, my head fallen back like a dancer, like a statue we revered in the Luxembourg Gardens in Paris. And when I looked up, the same reverence was there in your face just to look upon me. After, we breathed the same air back and forth into each other's mouths, lungs, bodies. We tattooed each other with a ballpoint pen. You drew a design with *I love you* in Bulgarian at the end of my Pisces constellation tattoo, on the underside of my forearm. I drew a design of heart shapes on one of your round butt cheeks. We took pictures of our tattoos, knowing they could never be permanent. Knowing they could never stay.

THE HOUSE OF PLEASURE

Blunt

Undermining your assertion that you would never date a white person again, undermining my assertion that I would stay single until I healed from the previous fall, we held hands a lot. Later, a half-smile on my mouth, a question in my eyes, I held an umbrella of yours that looked like the handle of a shotgun.

"That's the umbrella that almost got me killed," you said. When the police pulled you over for a blown-out license plate light, they asked for your ID and didn't explain why. Just probed the blackness inside your car with industrial-strength flashlight beams, searching without a reason, without a warrant, searching with the hope of finding something, anything to justify unleashing coiled violence. Illuminating their bullseye, a shotgun handle, they drew guns and pointed them at you. You held your hands up and froze. When they removed the umbrella, you had no words and neither did they. Your hands still up, you cried. In lieu of an apology, they lowered their weapons.

You want words to be precise, sharp with meaning. You don't want me to tell you how I feel with clichés. Instead of those three words used the world over: "My whole life," you say. All I can come up with is one curved, barbed, razor-edged letter.

"We are," I say, and we shorten this to just "are," then, "R," signing notes to each other with only this. And we know what it means. We know that we're naming the ineffable pull between us with more precision.

But sometimes I want to hear "I love you." I want the weight of history that comes with the meaningless. I like to blunt the sharp edge of words. I like to make them tickle, maybe sting, and even when I press them hard into you, I don't want blood on my hands. But I was born with blood on my hands. I wish I could find words to tell you how hard I've tried to make my hands clean. It's occurred to me my motorcycle accident was a metaphor, an attempt to tear my hands completely from my body, but I only got one off and the problematic skin that covers me is what held it on. As it hung there by this skin, I saw myself turned inside out and the colors were different. What you understand better than me, because you're a twin, is pain is the twin of pleasure. And so, it should come as no surprise when I next felt so inside out is when your fingers moved inside my body, bringing me out, and mine, long reattached by then, moved inside yours, and out we both were.

And what if metaphors are sharper and the words themselves can be softer? You like to blunt them sometimes too, because we both know; it wasn't the umbrella that almost killed you.

Bright Brain

Twice in my life, I have seen a dark, star-strewn tunnel with a light at the other end, toward which I was hurtling. Neither time I experienced this hallucination was I having a near-death experience. Once was from pleasure, and once from pain.

The first time, it came during an extended orgasm, which was the finale in a series of them — six or seven — produced by my lover at the time. Never before and not since has anyone produced in me so many orgasms all in a row. It was an extramarital affair approximately eleven years ago inside a mediocre hotel room in Phoenix, and I was in love with her and wanting her to leave him, but they had two small children. During that grand finale, my field of vision went black and ever so briefly, in a flash, I saw it. Over too soon, I lay in a spent heap with limbs entwined and stardust settling behind my eyes like a resting snow globe after one swift, hard shake.

The second time I saw this tunnel was while giving birth. They had transferred me to the hospital from the birthing center after they had manually broken my water and seen meconium in it. Often, babies can't breathe unassisted in such cases because they will have this sticky black excrement clogging their airways. This was an eventuality they were not prepared to handle at the birthing center: a six-bedroom old Victorian house with the rooms converted into birthing suites. When I first arrived there my contractions were coming fast with almost no breaks between, and my midwife explained it was because the baby's head was tilted against my cervix. She suggested I do my best at an inversion to help the baby adjust positioning and straighten out. The closest I could get to inverting was to get down on all fours and lower my face to the ground with my butt up in the air. We stayed in the waiting room for hours, as the day outside was hotter than tolerable and all the rooms were occupied by delivering women. From upstairs in one of the rooms, I heard a woman scream as though she were being murdered.

Hours later, I was submerged in a tub of lukewarm water, the taut mountain of my stomach protruding like a pale pain island delivering unrelenting agony, my legs open, my voice moaning low, my body a wounded animal. The pain was like no other pain I'd known. I can only describe this way: it was as though roots of a tree had grown thick and entwined around all my internal organs, and then a great, muscled hand grabbed the trunk and yanked and yanked continually for twelve hours until it all finally came out.

The tunnel came during the penultimate push, where my son's head came close to the opening of my vagina, and from his passage broke off the tip of my tailbone. I heard the crack and felt a short burst of white-hot pain at the base of my spine. The burst was so intense, I momentarily passed out. My field of vision went black and there it was again, that bright, dark tunnel punctuated by starshine.

///

When I imagine a visual of the conjoined twins that are pleasure and pain, I see a snake swallowing its tail — the overlapping beginning and ending of a circle where one extreme turns nearly indistinguishable from its opposite. In the early 2000s, several studies showed both pleasure and pain light up the same area of the brain. Further studies found dopamine, a chemical long associated with feeling good, is produced by the body in both pleasure and pain contexts as well. Because the only way to study the brain of a living person thus far is through imaging, which reveals blood flow in the brain versus individual neural signals, there is a paucity of knowledge about the exact mechanisms of this command center of the body. The current theory is when a stimulus is registered at the surface of the body, be it a bee sting or a kiss, signals are sent to the brain for processing and the brain decides how to react. If pain is the response to the stimulus, the brain also releases endorphins in the wake of that, which both block the brain from sending pain signals and flood the brain's limbic and prefrontal regions, causing a sense of euphoria. There is a heightened point in both sensations where it's indecipherable which is being felt, pleasure or pain, and that, perhaps, is where the consciousness is thrown into that starry tunnel. I call this state, Bright Brain.

///

As a prepubescent child, I used to bike with a group of boys. I wanted to be a boy myself. The boys rode bare-chested, t-shirts tucked into the backs of their waistbands and flying behind them like capes. I removed my shirt once I was certain I was out of eyeshot of my house, where any member of my family might have seen me. Did the boys know I was not one of them? Or did they know and not care? The pleasure of that feeling — my small, bare-chested body flying through the soft summer air upon two wheels surrounded by other children who looked just like me, all our chests the same, indistinguishable from one to another should just our torsos and limbs be photographed, our headless forms. I had been thinking of us headless — like a black-and-white photograph of a group of children lined up shirtless, pale chests all in a row, dirty jeans with ripped knees and gangly arms with prominent elbows — when it happened.

The group stopped abruptly, and I stopped. Not having the skill quite yet to grip the hand breaks of the new ten-speed bike, I slid off the seat to drag my feet and my crotch slammed into the straight bar. Even as water pooled then spilled from my startled eyes, I couldn't quite tell at first what I was feeling, pleasure or pain. There was nothing but strong sensation swirling through my body, round and round like ripples on the surface of a pond.

/ / /

The tunnel with the light at the end is proverbially associated with dying, yet this image always makes me think of birth. Wouldn't that be the experience of a new human coming out from inside her mother into the world? In fact, studies have shown the signal that starts a woman's labor is sent from the baby. Perhaps death is a birth into what comes next. Perhaps, the signal of readiness comes from the dying person. Birth and death being two sides of the same coin seems obvious. And the connective tissue — the entryway, the exit way, the tunnel, the vagina — is sex.

A theory that emerged in the early '90s proposed this phenomenon may be so frequently reported as part of near-death experiences because it's the effect of oxygen deprivation to the brain for vision to narrow and tunnel with darkness all around but light at the end. The theory came from British psychologist Susan Blackmore's 1993 book *Dying to Live*, which explains the phenomenon as a result of chemical changes associated with dying brains. Blackmore argues a lack of oxygen (or anoxia) during the dying process might induce abnormal firing of neurons in the part of the brain that controls vision, leading to the illusion of seeing a bright light at the end of a dark tunnel.

More recently, some non-materialist neuroscientists, like University of Montreal professor Mario Beauregard, have critiqued Blackmore and explicated brain research was in its infancy twenty years ago.

Beauregard cites objections by Dutch cardiologist Pim Van Lommel that if anoxia is central to this phenomenon, far more cardiac arrest patients would report such an experience. Also, as pointed out by Dr. Sam Parnia, whose resuscitation techniques have doubled his New York hospital's cardiac-arrest-recovery rate, some near-death experience patients were not terminal during their experiences, meaning their oxygen levels were normal.

///

Around age nine, I went to a nearby playground sporting my new bowl haircut accessorized with red rubber boots, a blue Izod sweater, and jeans with holes in the knees. Before I left my house, I looked into the mirror and smiled: fun hair, cool outfit, straight hips. Pleased, I ventured out, the sky swelling and gathering above me, thick, taut air holding it all in. There was a group playing keep-away around the slide, girls against boys. I joined in, climbing up the slide after one orange-haired boy, his bright locks bouncing. Arriving at the top, he turned and looked at me. He thrust down his palm and held it there above my head. I stared up at him, his small eyes nearly lost among a mass of fiery freckles. Confused, I reached up and slapped his hand. With a "whoop," he went over the top and swiftly disappeared. I climbed into position to go down after him when a girl down below pointed squarely at me, shouting: "There he is, up there! Get him!"

He? Him? I looked around, redly smiling, preparing to laugh when her mistake was discovered and pointed out by one of the other kids. I shivered. A voice behind me shouted: "Go down, man, run!" I swooped down, leapt to my feet at the bottom, and ran, settling into the queer power of how it must always feel to be a boy. It was intoxicating yet frightening to deceive these girls into believing I belonged on the opposite team. At any moment they would notice my soft, long eyelashes, a girlish curve to my cheek. I would move in some un-boyish way and give myself away. But when I high-fived the boys and took off away from the girls, they continued to believe I was a boy.

I ran fast, away from a dark-haired girl, looking back every few steps to see her still pursuing me. I tripped on a protruding root and fell, my knee slashed open on a piece of slate rock, my face hit the dirt. I lay there, splayed out on the ground with a bleeding knee and a scraped cheek. It began to rain. The girls screamed, everyone scattered. I turned over and found the dark-haired girl crouched next to me. Our faces, so close, seemed to freeze, eyes locked, her black pupils shiny, beckoning. A car honked in the street. A truck rumbled by. The gritty taste of dirt lingered in my mouth, the smell of extinguished campfires hung, trapped under the rain. She placed one small hand on my shoulder, tilted her head, brushed a fingertip across my scraped cheek. "Does it hurt?" she asked. I tried to speak but couldn't get my mouth to move. The rain fell harder; a voice called from a distance. She leaned forward, summer-scented, and left the soft tremble of a kiss on my wounded cheek. She ran off. I sat there, drenched, a smile so big it ached.

///

Those who nearly die also commonly report a sense of euphoria and/or a complete loss of pain. Similar phenomena, described more poetically by one user as "a whole body orgasm," have been described by people who use ayahuasca, a psychoactive beverage derived from plants that are rich in β-carbolines (harmine or tetrahydroharmine, among others) and N,N-dimethyltryptamine (DMT), respectively. This beverage has been used ritualistically in Amazonian cultures for centuries. Recognition of this heritage has led to the piecemeal legalization of ayahuasca drinking rituals in countries worldwide, including the U.S.

The pineal gland, also known as the seat of the soul, is the only structure in the brain that is not found in a pair. It is a singular and central structure that continues to mystify neurological scientists as to its functions. Some of its functions are thought to detect light and dark in the environment and set the body's circadian rhythm in harmony with that cycle. During the night, the gland will secrete melatonin to aid in the relaxation and sleep of the body. There is research suggesting its function is to be an aware and alert sentinel in the body while we sleep. This is the gland that will suddenly wake us if, for example, an intruder climbs noiselessly through our bedroom window to attack us.

Descartes believed the pineal gland is the source of all thoughts, though we now know that's not true. Neural networks are that source, and such networks can be found throughout the body, including around the heart and in the genitals, but most by far are found in the brain. When I learned of these bodily neural networks, I felt a suspicion I'd long held was finally confirmed by science — the body not only stores and holds memories but produces thoughts. It's now scientifically sound to assert the heart, in the midst of the phenomenon of falling in love, may possibly tell the brain what to think.

///

Unwilling to wait for companionship, tired of the peculiar loneliness of being single, I go to Tinder. I know I will most likely only find hurt people pursuing more people that they can in turn hurt, but I try to keep an open mind. Two of my close friends, after all, met each other on Tinder. They married and had a baby, and they are living their dream in Amsterdam, and it's been ten years. If it could happen for them, why not for me? And yes, there are a lot of hurt people out there, but not all of them are stumbling around blindly hurting people like these last few I've inadvertently coupled with. I'm not doing that. Awake and alive and conscious of my hurt, and conscientious of how it affects my behaviors in intimae settings, I take care not to cause harm.

And if I'm on here, surely, there must be at least a few others like me on here.

I swipe right only on about one in twenty faces, and after about twenty minutes of this, I get a match. She texts me a heart emoji. I reply. We meet for coffee, and I feel a sense of bleak futility sitting across from her. The air is cold, and she is underdressed in shorts and a t-shirt. Her hair is long and black, and her eyes are brown and penetrating. She makes deep eye contact, which leaves me more quiet than usual. We banter with small talk for a bit and then she says something that surprises me: "The world is a menu; you pick what you want."

A waft of her scent rises from her body and rides the air into and across my face and it's a sweet and floral smell, like candied roses. Pleasant and maybe too much. The combination of the look of her, the smell of her, and these surprising words she uttered prompts me to ask the question.

"What do you think happens when we die?" She opens her mouth to speak, and I don't listen. I hear sound, and she is making words, but I can't attend their meaning. I'm too focused on the way she is filling two of my other senses — what I see and what I smell — and whether it is too much or not enough.

'What do you think happens?" she asks.

"This is a topic of great interest to me, so I'm glad you asked. My answer is, I don't know, but I have theories, and I'm 91% certain about the first thing I think happens when we die."

"Which is?"

"A lot of pleasure," I say, and start to say the next thing.

"DMT," she says, interrupting me. I'm stunned into silence.

"How do you know that?" I finally utter. I had done a bit of digging to find out there is a relatively new theory around this Bright Brain phenomenon that happens with death and how it is caused by DMT. It certainly isn't common knowledge, and the fact that she spurted it out so casually off the top of her head stops me in my proverbial tracks. The seduction tracks. The ones I hopped on at some point in the conversation with the goal of stirring desire in her. Well-researched tracks built and adjusted over so much time and experience with the game of seduction, moving me in one direction, the one that works best.

"I read about it somewhere," she says, "and it happens in women when they give birth also."

This surprises me so much I now feel derailed. What is this glitch in the matrix? Who is this woman from Tinder? She smiles at me, her brown eyes filled with something that looks like knowledge only she possesses, and I do not. The wind blows her hair into her face and carries her smell again to mine, and it's over too soon. It's not enough.

THE HOUSE OF RAGE

SINGER | Heavy Duty

Stay Hurt

"What's the worst thing you've ever done?" I ask you, my twelfth love. And you answer something mild and forgettable, something like, "I yelled *shut up* at someone once." Only now, you'll have a great answer for that next time, won't you? Or will you lie? Were you lying to me?

I can't indulge in victimhood, not really, because when I was eight, I stole four cans of beer from the refrigerator, took them out into the woods, drank them all in quick succession, and buried the evidence. Intoxication became my elusive solution, one not readily available to a child. Until my sister had a near-fatal car accident when I was ten. Then all supervision vanished, along with all forms of attention. Stealing was my way of providing for myself. Whatever I wanted, I took it. Candy and cookie dough and alcohol from stores. Money from my mom's purse. Random possessions from the desks of my teachers in grade school. In middle school, I set the Home Economics room on fire, burning the plastic covers on the sewing machines. I still remember the chemical smell of blackened, melted plastic. Some boys falsely took credit for the crime, wanting to enhance their reputations. I was never caught. And I would go on to do worse things.

You're dating a man and wanting to date a woman on the side. I'm dating a nonbinary, polyamorous millennial with a large intellect and an even larger vocabulary who has convinced me monogamy had been my problem all along, and nonmonogamy is the solution. I primarily started dating you to artificially increase their desire for me, as I noticed when I was paying attention to a second love interest, they liked me more, and wanted me more. I felt sure there were some red flags, but there was one scary yet thrilling piece about their philosophy of nonmonogamy I was compelled to try. And yet, I was afraid it would all not work out well. Because even though it looked good on paper, I know communism does too. But when I travelled to Russia at age seventeen and saw the devastating aftereffects of communism — including people with open wounds living on the street with no access to healthcare; an orphanage with wall-to-wall cribs filled with crying, unheld and unloved babies; and crumbling apartment buildings that looked like concrete cubes stacked atop one another — I knew it wasn't good in practice.

"I think I'm addicted to you," I say when you come to spend the night with me that first time. "I feel high around you."

This makes you smile. But then you say, "I hope that's not true."

"Why?"

"Because then you'll have to abstain from me."

This first night we spend together, I notice your feet, and how they seem to have plastic stuck to their bottoms. You have flat feet, you explain, and what looks like plastic pieces are actually huge calluses on your soles. This is a clue I do not see. A clue you are not soulless, but something has hardened around you, closing you in. I know you have a dark side because we all do. I expect your darkness to be smaller than your light, but as dark matter makes up most of the matter in the universe, so it is with you.

When I ask you that question about the worst thing you've ever done, I do so with the desire to confess the worst thing I've done, and it was worse, maybe, than what you and he did to me. I figure I should be able to forgive you because I've done worse and been forgiven. Specifically — technically — I raped a woman.

///

The worst thing I've done happened half a lifetime ago with a distant ex-girlfriend, my third and my fifth love. We were in our twenties. I was sober and she was a relapsed alcoholic. She had been my girlfriend, but she wouldn't sleep with me. Strangers at bars, she was happy to sleep with, but not me. When I asked her why, she said, "Because I'm in love with you." It didn't have to make sense. Her body, her choice. But then, it did make sense — love makes people feel vulnerable, and sex while vulnerable makes intimacy, and intimacy is, especially in this context, terrifying.

I realized I could get in the way of the strangers if I went out to the bars with her. I drank Diet Coke while she drank shots. I picked her up when she fell off the barstool. I took her home, tucked her into my bed, and slept on the couch. Night after night, I did this, until the last night. "I'm moving back home," she told me, "where I can get sober again." We went out one last time on her last evening with me, the night before she flew away with a garbage bag full of her clothes.

///

Your kindness is a balm. It isn't until later, when my first partner in our poly equation is gone and my time is freed up to spend with you and I ask you to be monogamous with me, that I realize for almost ten years, I've been with women who were quite mean and cruel to me. This included the nonbinary person who was assigned female at birth, and who, during our relationship, came to me almost daily with a new list of grievances.

"I need you to do better," they said with a stern face and that perpetual disappointment in their eyes on a video they texted to me, after I texted a video of my dog doing a trick. The trick, which my co-parent taught the dog, was to mimic a gun going off with thumb and forefinger and shout "Bang!" and the dog would fall to the floor and roll over, wagging her tail in anticipation of the reward of a treat. This had been offensive to my first partner, and they wanted to know how I could be so thoughtless as to send it to them. But they loved my dog, seemingly even more than me, and my goal was only to make them smile. And yet, I was used to being wrong when my intention was to make someone smile. I'd spent a decade of being refrigerated downwind between passive aggressive and overtly aggressive partners.

Your kindness wraps me inside of it and warms the chill in my bones. You listen well. You are thoughtful, supplying my favorite things in your apartment when I come to visit — Philz coffee, organic half-and-half, even my own mug with an *E* on it that was my favorite color yellow. And you are an excellent gift giver. You also seem to love your friends and family and are clearly loved deeply in return. And your friends are all old friends, with you since childhood or at least for a decade. You're generous with your time and energy, talking with me on the phone every night, sometimes for hours, and many times driving for three hours to spend as many in my bedroom with me after my kids go to sleep. And all this adds up to a high-quality person in my estimation. You say you don't know if you believe in Reiki, but when I touch you, you say the palms of my hands burn.

///

I got to talking with someone at the bar and lost track of her. When I went to find her, she was kissing another girl in the women's restroom. I called her name and felt the woundedness on my face, which made her laugh. She grabbed my hand and pulled me out onto the dance floor and kissed me and groped me. She tasted like whiskey. It was flashing lights and pelvises pressed and hands handing and tongues tonguing and the song was "Die Another Day" by Madonna and I was twenty-five and I didn't know. I had no idea that five years later we would be living in a house together in a place she called home with two dogs and two motorcycles and two cars and she would give me a book for Christmas that she had written and illustrated. In it, she would have drawn herself jumping off a cliff, and it would be captioned: *Me if you ever leave me.* And then, less than a month later, she would abruptly leave me, and she wouldn't even think of jumping off a cliff, nor would I. But I would circle a gun store because the place she called home has gun stores, and I would fantasize about parking the car and going in and purchasing one and shooting myself. But I was twenty-five and I didn't know.

From there, I took her home again, but this time I didn't sleep on the couch. She took off her clothes and I took off mine and she lay on top of me on my bed. It hurt when she put her finger in me because my vagina was dry. I entered her the same way, and she didn't respond but to drool on my face. It was in that moment it hit me hard: *This woman is barely conscious. She does not know this is happening, this event she explicitly stated, while conscious, that she did not want to happen.* I removed my finger from her vagina and wiggled my body away from her and turned over. She immediately started snoring. I pressed my face into my pillow and cried.

/ / /

"I'm sorry," I say to you, my twelfth love, after breaking up with my first partner and trying to be just your side piece for a few weeks before running into a wall of hard jealousy. "Polyamory isn't for me. I thought I could keep doing it, but I can't. Let's be friends, though, and if you ever feel like you want to date only me, we can talk about that."

When you say you want to date only me, and you will break up with the guy, I respond you don't have to. I assert you should only do it if it is what you honestly want.

You lie to me and tell me it is.

Six months later, when I sit locked in the bathroom of our rented house while we're away on my birthday trip, looking at your phone and reading your texts with him, the evidence is clear: You had merely used me to turn him on. To get more attention from him. What stands out is an audio recording he made of you, the sounds you made, and he commented about it, and you replied, *Maybe it was all the anal.*

At that point, I come out of the bathroom to find you sitting on the bed looking anguished. "It's pretty bad," I say, having seen you texted with him all day, every day, and he sent you pornographic images of women and you responded with what you wanted to do to them. And in addition to that game, you gave him a blow-by-blow of our relationship, our sex, the dildos we used (you had sent photos of them laid out on my bed), even images of me without my clothes on. Images of us naked together. You made him believe I would potentially join the two of you, and he could have sex with me, too.

You did whatever he wanted. And he wanted it all, including me. And you? You gave me to him. This is never something you admit, but I know it's because you are too kind. Even right up to the edge of responding to my ultimatum by giving me the evidence, the truth, you lie. I see how you just don't want to say anything that will hurt. And when you know the truth will hurt, you lie instead. Another kindness.

///

I don't remember sleeping that night. When the morning light pierced the room, I rose and put on the same clothes I'd worn the night before. Smelling like a stale cigarette, I went to the kitchen and made coffee. A while later, she walked into the kitchen and sat down. Her eyes looked like blown-out candles with little coffee-stained pouches underneath. She wore a cheap white tee, the kind that comes in a package of three at the gas station, and there were ash stains on it, and her bloated torso beneath the sports bra she always wore resembled a small barrel. For a moment, I saw her the way she really was — puffy and haggard and sick with alcoholism. Suddenly, there wasn't that soft blur, the one that makes all surfaces gleam with perfection. Her dark brown hair was short in the back and long in the front and when she moved to pick up her coffee and sip it, it fell over her eyes, and the soft blur returned and she was perfect. I straddled her lap and wrapped my body around her. I felt her arms around my back. I started crying again, and I didn't stop when she got on the plane. And I didn't stop when I arrived back home. And I didn't stop until late that night when I was finally too exhausted to cry anymore.

///

You had presented your other lover to me as a friend, and had me meet him, talk with him, even endorse him as a person. Meanwhile, behind the curtain you had drawn over the truth, you were telling and showing him every intimate detail about us. You even graphically described my vagina to him (I was impressed with the writing). The worst part, though ...

After I tell you it's pretty bad, I sit on the bed beside you and scroll faster through the endless messages. Miles and miles of scrolling, and I wait for you to say something, anything in your own defense. "What?" you say, as if you don't know what. I stop scrolling and see a deleted video that you had sent to him. "Did you send him our sex video?" I ask.

"No," you assert, reaching for the phone. I pull it away and look at the context.

You: *Do you want to watch it on your own or wait until I get there and we can watch it together.*

Him: *Whichever you want.*

You: *OK, go ahead and watch it, but please delete it right after. Promise? She's in it too.*

Him: *Oh God that was hot. OK, reluctantly, I'll delete it, but I really don't want to.*

"Yes you fucking did send it to him!" I yell and shout-read the words back to you.

"I guess I did then. I didn't remember, I'm sorry."

I throw the phone down on the bed and feel the panic and rage rising in my chest. I curl up in a ball on the bed and hyperventilate and cry. You wrap your arms around me and apologize over and over and all over.

/ / /

Four years later, I moved to that place she called home to live with her. By then, she'd been sober again for two years, and after we had sex that first time (real and wet and drool-free and everyone capable of consent), she looked at me and smiled and said, "I'm so glad we didn't do this back then." The silence after the words came down on me with the weight of mountains and I was crushed. My heart raced and I felt the woundedness on my face, which confused her.

"We did, remember? That last night after the bar?" She laughed then and said, "Oh god, I'm sorry, no. I don't remember." I didn't say anything more. If her perspective was that ours was just another in a series of drunken and forgotten sexual encounters with people at bars, then I was absolved. My estimation of my trespass was wrong. But I had not been drunk. And I knew I was wrong.

Eighteen months later, she cheated on me and left me, and I couldn't help but wonder if it was her way of punishing me because, really, she knew. I was the sober one. I was the responsible one. I was supposed to be keeping her safe. I wasn't supposed to prey on her weakness and take advantage, like those two older boys did to me when I was fifteen, and I came to from a blackout naked with one of them standing on my hands and the other one on top of me, taking his turn. And it would be all of this — her cheating and lying and leaving and punishing and what I did and what they did and all the ruin — that would have me circling the gun store.

///

That video was supposed to be private and just for us. You know that. He knows that. He solicited these things, knowing I wasn't aware, much less consenting. This man whose character I had mistakenly endorsed was, in fact, a misogynistic piece of shit who intermittently used and ignored you, and whom you chased and doted upon because of your own old wounds. This is only clear to me because I've danced with very similar shadows. No woman raised in the United States can escape internalized hatred of women, nor the harm that necessarily weaves through our psyches, bodies, desires, exploits. It is this, your desire for toxicity, which is the root of your self-loathing, which shows itself in brief flashes — saying you look ugly without makeup, calling yourself fat, warning me nobody ever likes you for long — but which never made much sense to me until I flung open that curtain of duplicity.

But that's not what I want to address here. Not the facts that led to our undoing. Instead, I want to write about the mysterious phenomenon I believe I experienced with you. Mysterious because it has no name, no form or formula. It's like a dazzling golden tendrilled light just at the corner of the eye that disappears when you turn your head: You can't see it, but you thought you could, and you can still certainly feel it. You feel it viscerally, starting low in the abdomen. A hollow ache fills you. Its effect is to make all sensations involved in romance more vivid. Saturate the colors, increase the exposure, apply that soft blur. Its nameless presence vibrates in the air like magnetism but intensified. I name this mysterious phenomenon, The Vivid.

I believe The Vivid is responsible for my outsized appetite for sex. The sex I crave is not just any sex with any other warm body, nor will pleasuring myself sate it. (On my own, I prefer a good sneeze to an orgasm, most days.) In fact, I have long asserted what I see now, was a falsehood. Remember when I told you the story, too? (You loved my stories, and I loved that you loved my stories.)

I told you I didn't like to have casual sex with women, and in fact, am not capable of it. I can have casual sex only with men. I explained how I'd learned this through trial and error, and it proved itself out more than twice. When I have sex with a woman whom I find objectively attractive, but with whom I share little to no emotional connection, it feels gross. When I have sex with a man with whom I have an emotional connection, it feels gross. By gross I mean, like I'm sleeping with a sibling. However, when I have sex with a woman I do feel emotionally connected with, I enjoy it. And, strangely, have, on the off occasion, found it fun to sleep with a man I barely know. "Paradoxes," you say, "happen when both opposing concepts are equally true."

Now, though, I see I've been missing something key. I can, in fact, have and enjoy casual sex with women, too. Because sometimes there is an emotional connection, and there is newness, and these can masquerade as The Vivid. In retrospect, I can easily spot the differences. Like the fundamental forces in our universe (gravity, electricity, magnetism, and radioactivity), it's not necessarily subtle. Its presence is unmistakable. But sometimes, briefly, a bout of vertigo can be mistaken for radioactivity, or a jolt to the funny bone for electricity. And, on the occasions I mistook newness for The Vivid, my delusion didn't last, but the relationship did. I stayed in it, each time for various reasons. And I continued to have sex with that person and enjoy it. And, outside of the context of The Vivid, it was decidedly casual.

I have a skill for weaving logic in any way that will support the narrative of the moment, and I am the writer. Then something simple pokes holes in my carefully crafted logic, like the way my dog looks at me. And how she looks at me this way every day I'm with her, even if I leave for a week. It's a depth of sight and a way of seeing not often accessed by human animals. To describe it as pure, accepting, or even unconditional — these would not be right. Maybe it's better described by what it's not: judging, calculating, vengeful.

You never much liked my dog.

///

For years, I didn't speak to her. Then, there was a nationally televised hearing on the floor of Congress because a man — naturally a white man of great privilege — had been nominated to the Supreme Court but a pesky bit of his history had emerged. An incredibly brave woman came forward to tell the nation that when she was only fifteen, this man had pinned her down and attempted to rape her. Watching the hearing made my stomach hurt. I thought of them, the boys who raped me when I was fifteen. And I thought of her, the girlfriend I raped about ten years after that. For a while after fifteen, I didn't care anymore about sex or consent, and I lost count of the number of times a penis was shoved into my mouth or between my legs without my permission. I couldn't be anything but prey, and I couldn't stop the predator with a thousand faces, so I internalized him.

/ / /

The fundamental forces share one feature: their unmistakability. But of what are they made? And what about hidden dimensions, dark matter, consciousness itself? Quantum principles arose to correct and fill the gaps left by classical theories of physics, gaps that centered around life yet not accounting for life, this bright phenomenon. Scientists discovered quantum physics when studying the light emanating from hot bodies. From your observable perspective, objects in motion tend to stay in motion, those at rest will stay at rest, and the closest distance between two points is a straight line. Bodies in space move in perfectly predictable and uniform patterns, and in such ordered predictability, you find a sense of control. A sense of comfort. But then I came along, and you experienced something new, something you had never previously known to be true: There is no such thing as rest; there is motion inside matter; there is actually no perfectly linear nor perfectly uniform motion. Motion that is faster than the speed of light exists in the microscopic domain, and our bodies have always taken advantage of this. Consider the speed of processing a surface stimulus, just for starters. For each visual input, it takes a tiny but perceptible amount of time — about two-hundred milliseconds, one-fifth of a second — for the information to travel along the optic nerves and into the brain to be processed and interpreted.

Perhaps The Vivid is a product of one of these hidden dimensions, unknowable by the human animal except via this mysterious effect of two particular bodies in proximity. Consciousness — the aliveness in and around the human animal — itself has dimensions, seen and unseen. The Latin root of the word, conscire, means to be in the know, to be privy to. The root word then evolved into *conscious*, which means knowing *with others*. The addition, subtle yet important, opens the door to a new vista of understanding about The Vivid: it exists only as a shared experience. And this hidden, unseen, shared experience dominates my desire nature. To have The Vivid is to be inspired, to feel alive, to radiate beauty and joy like a stone dropped into still water; though swallowed instantly by its depths, the effect of the stone's weight ripples far and wide along its surface. Similarly, dark matter and dark energy are theorized to comprise up to 85% of our universe. Their presence, though invisible, is measurable by their effect on the observable universe. Dark because unseen. Dark because unexplainable, at least completely, by any existing theory. Dark because, as always, more is unknown than known. In fact, everything on Earth, everything ever observed with all human-fashioned instruments, all sensible matter, adds up to less than 5% of the universe.

/ / /

The entitled white man on trial for sexual assault got to be a United States Supreme Court Judge anyway, because he said he didn't do it, and his side of the story mattered more. What followed though was an uprising; woman's voices shouting, "Me too!" and naming their predators and rapists and assailants. I knew I couldn't begin to name mine until I first named myself. I reached out and asked her if I could come back to that place she called home and visit her, and make amends. Trembling and speaking through a boulder lodged in my throat, I sat beside her in a public park and told her everything. When I finished, she looked at me in a way that reminded me of how my dog looks at me, put one of her hands on my leg, and with kindness in her voice, she said, "It's OK."

I wanted her rage and her wrath, I wanted her to send me to prison, I wanted to be punished by her forever. But all she did, so quickly and with three small words, two of them conjuncted, was to forgive me.

One of them not a real word, but two letters, one eternally round and wide open, allowing for it all. The other hard and pronged and liable to cut. The origins of the *not* word are unknown, but not unguessed at. The same could be said about the origins of my not-choice. I did not choose to do any one of the other myriad actions or nonactions with my clothes still on that I did not do that night, stark raving sober. My not-choice was my clothes coming off and my naked, conscious body being positioned under her naked, barely conscious one. My not-choice was her drool hitting my face like a bullet. My not-choice is a haunting. It is a fanged phantom trapped in my body. It is a hovering at the corner of my eye that, when I turn my head, is not there. It is both the apex and the nadir of my story's arc. It is not sex. It is not death. It is both.

///

Dark energy, the strange energy-fluid filling the universe, may have something to do with The Vivid. It may flow into and through The Vivid, an attractive force not easily resistible by the human will. My most powerful attractions are inexplicable. Unlike my friend Julia, who can run down a list of compatibilities, boxes ticked off one after another (educated, fit, conventionally attractive, wealthy, intelligent), I usually have less in common with those I'm most drawn to than anyone else.

Perhaps then, The Vivid is just another addiction. A pathology. And when I feel addicted to someone, I tend to romanticize it, make it mean something magical, like fate or beyond-death-significant. Every feature turns adorable in my eyes. Every gesture is sexy. Her hands on my skin feel electrified. And it will surely always feel this way. Time and repetition won't blunt anything here. We will never take any of it for granted, ignore each other, or feel bored. We will always only rage through life together with hearts on fire. Maybe The Vivid means I'm addicted to her and using her like a blank screen upon which to project this impossible fantasy. And it only can work to use her if she's also using me. And the next layer I'm less aware of is how maybe some broken part of me believes it could get healed by making her shift from using me to loving me. Maybe if she devoted all her love to me, my self-neglect and self-loathing would either disintegrate on contact or shrink to such an insignificant size that I would never notice it again.

///

The day after she forgave me in the public park, I visited her house with my son, who was then about four years old. She had baked apricot bars for us because she remembered how those had been my favorite back then, in the time of the two dogs and two motorcycles and two cars and our house. I watched her watch my son with wonder and delight on her face. "He looks just like you," she said, and I thanked her, considering it a high compliment. She had a cat who my son was chasing and who wanted no part of that, so the cat reared up and hissed and scratched my son's hand. Bright-red bloomed on his finger, and he cried. She ran to fetch Band-Aids and I found a tissue to press against it. When she returned with a Band-Aid, I started to put it on him, and he pulled his hand back protectively.

"You don't want the Band-Aid?"

His small face crumpled and hardened, and his small chest puffed out. He held his tiny wound, cradling it with his other hand like a cherished pet, like the one he'd just been chasing but couldn't catch and whose pursuit had left him with this. This wound. This victory. This badge.

"No," he said, "I want it to stay hurt."

///

You want me to forgive you and give you a second chance. You will be all in, you say, and you won't talk to him anymore. You didn't realize how bad it was, you say, until you saw it through my eyes, and you now see how wrong you were, and nothing like this will ever happen again. You ask if I will give you a chance to make it up to me. In one of the nights we spend together in the couple of weeks after, I have a nightmare. In it, I am naked and drugged and unable to move. You are there, and you have let some men into the room to see me naked in the bed. They have sketched my body and you are showing me the sketches. Angry, but lethargic and unable to express my anger, I ask why you have let them in. Why did you expose me to them this way? You scoff and tell me not to overreact, then you take the sketches and leave the room. Soon after, a man comes into the room and sneers at me. I ask him what he wants. I tell him to leave. He starts taking off his clothes and laughing. Terror fills my body and I writhe around and scream. I wake screaming with sweat on my back. You are there beside me in the dark. You tell me it's OK. It was only a dream. You hold me.

In the end, I break up with you. I try, but I can't forgive you and I don't see how I will ever be able to trust you again. I wanted to forgive you. I wanted to put a hand on your leg and have kindness shining in my eyes and genuinely say to you *it is, it was, it will all be, all, OK.* I wanted to look at you the way my dog looks at me. Although, I don't think it's a conjurable look. Trying to look at you this way would yield something altogether different, kind of like the gap between what I sometimes want to say and what I end up saying when I open my mouth. But I can't make the sound *O* makes without opening my mouth. And I could never have received that generous forgiveness I did not deserve with my eyes closed.

The callouses that look like thin pieces of plastic? They're metaphors. There are key emotional contours you also lack, and the repeated pressure, the bearing of emotional weight has made thick callouses that prevent you from sufficiently feeling what's true. You can only feel it if it's hard, sharp, cutting. You can only feel it if it hurts, and only if it hurts more than you already, always hurt. You can only feel it if it's false. And I don't have time for false.

And this phenomenon I call my most precious resource — time — is itself a darkness. Where does time exist? Only in space, and only relationally. In fact, just as The Vivid exists only as a shared experience, time only exists as a shared experience. It's merely a way to measure bodies in space moving in relation to other bodies in space. And likewise, The Vivid is a way to measure the degrees of intensity inside the invisible texture of a strong attraction. Or perhaps, I suspect, it's a way to measure the degrees to which I'm holding onto, like a cherished pet, deep-seated pain, wanting it to stay hurt.

You asked me where the idea came from, that one day in the afterlife is equivalent to one lifetime here in our animal bodies, stuck to the surface of a spinning ellipsoid, hurtling through space. I don't remember how I answered, and I don't remember the source of this idea. Perhaps it downloaded from the cosmos into my central nervous system just as I was about to die. Perhaps the act of dying opened a portal to the cosmos and all expanded wisdom available to us after life, after death.

Instead of leaving you, or even instead of trying to forgive you and staying, I wish I could have faked my death, so I could pretend to assume omniscience and know what you were really thinking, whether your intentions were benign, and if you remembered me fondly. All you had to do to remain so mysterious was hold silence. But from a point of view outside this space and time realm, you wouldn't be so mysterious. I would know about all the performing and faking and pretending you do.

Even though I don't know what The Vivid is, and regardless of what I suspect, it doesn't change its effect. It doesn't change how all my cells brightened at the sight of you, the smell of you, the feel of you, the sound of you — all my cells turning face-first toward the light like tiny sunflowers in a vast field. Likewise, as with sunflowers, they bowed their heavy heads at night to sulk face-first into the dirt. What that really means, you should know, is that sometimes with you, I faked my orgasms.

It seems I can't write my way out of not forgiving you. I'm not here to punish you, and maybe, you could never be punished enough to satisfy me anyway. What if forgiveness starts with self, and I have to go back to the child I was and forgive the secret arsonist me, the stealth thief me, all the way up to and beyond the sober young adult trespassing me? What if forgiveness in this context is not a choice? What if that not-choice was an undoing, an impossible going back and doing a different way? A way where you reveal the painful truth, unbidden? And I put my hand on your leg, and look at you the way my dog looks at me, and forgive you with an open sound and a sharp sound? But this story has already been written, and I can't unwrite it.

If I were a better writer, I would write a door we both could walk through and close behind us where these events never played out, and where our bodies could be erased of their histories and made new. Not healed, but unhurt. Like newborns, but full grown.

Here, Again

I have lived and died an alcoholic so many times. Getting sober — it felt like an experience outside the purview of my brain. My twenty-three year-old brain whose development was disrupted for thirteen of those years because of the alcohol I pickled it in, making me emotionally ten, and about that in maturity, too. Walking from my car to the door leading to the church basement full of cigarette smoke and coffee, I dropped my gaze to my shoes, to the cracks in the blacktop, and it all felt more familiar than possible. A feeling like I had moved through this exact moment in space and time before but in some other form. I stopped walking and looked up. The tableau of faces, the clothes on the bodies, the colors of the cars parked in the parking lot, even their shapes — all of it was known to me. Bright-yellow and child-blue, the curl of those fingers around that mug, the sideways glance followed by a full-on stare, the eyes behind that stare, and the eyes behind those eyes — it lasted longer than usual and I knew: I have lived and died an alcoholic so many times.

The people, that place, the whole thing I was there to try to do — it was making me feel one-hundred years old. My legs threatened to collapse, because they had supported too much of my weight for an entire century already, and how could I expect them to just keep doing this? Just keep holding me up? Keep moving me forward when the only place I was going was in a devastating circle that gets smaller over time, that closes in on me until here I was again? And again. And again.

I kept walking, kept moving toward the door while wrenching my eyes away from everything too old, too familiar, too the same, but finding only more of all that. How could déjà vu be so extended? It's supposed to be fleeting and over almost before it begins. Over so fast you can't be sure it happened at all. The effect of it fading as surely and quickly as the memory of a dream upon waking. Close to the brick church building then, dusk descending into darkness around me, I leaned one shoulder against the wall to steady myself on my elderly legs, my legs so unreliable, less reliable even, in that moment, than my actual memory, which could not possibly have extend beyond the short life I had experienced up to that point in this skin, on this earth, the unfriendly place I found myself. Suddenly I understood: I would not get out alive. This time was it: I knew I had to move through it to the other side and die sober.

Did a bird circle overhead or was it three birds? What did their circling signify as the bad news I didn't yet know not sink in? At the time, all I thought I had to do was die sober. All I had to do was die during an interval of non-intoxication. As impossible as even that seemed in that moment, it was far more possible than the actual truth, which was I had to heal first then die sober. I had to heal not just from my twenty-three years of wreckage, not just from my emptiness and sense of worthlessness at the time, but from many hundreds of lifetimes of the same. Many thousands, even. A degree of wounding that amounted to all the wounds of all humanity throughout history, combined. My brain was there to protect me from this understanding, which, had it occurred to me, would have crushed me under its weight. Where had this soul been to? How many bodies has it inhabited? How many of those bodies had been trafficked by misfortune, not survived, battled addiction and lost that battle? These were not questions my brain would let me ask much less theorize answers for. They were questions for later. For a more expanded time.

The growth and expansion akin to healing may have started right there in that church parking lot with its familiar cracks, shoes, colors, faces. It may have been gradual and slow. I may have locked my keys inside my car while it was running dozens of times after that and had to call the local police to come break into it for me. This may have lasted for months or even years, years during which I may have not felt any growth or expansion but instead, at times, its opposite. I may have continued to hate myself, starve myself, subject myself to unwelcoming spaces and unloving hands. I may have continued, for a long while, to lie and steal and be unhappy. In spite of all this, I may have, by some unknowable miracle, not picked up a drink or any other mind- or mood-altering substance. And in doing just this not small marvel of a feat, I may have, in spite of myself, grown, expanded, healed enough to get to a place where it no longer felt natural to lie. It no longer felt tolerable to steal. Not because I was afraid of my own or other people's judgement about those misdeeds, but because they became incongruous to my new nature.

Getting well inside of the unhealed filters of the human condition made for a stained and unpretty, wrinkled kind of self-regard. On my surface, I may have looked no different. Had they looked closer though, they may have seen the rage between my fingers where, in my nightmares, I murdered men with my bare hands by pounding their faces into concrete. They may have seen how this rage was right and necessary to the process. The healing would never be as full or anywhere near complete without it. Every injustice had to be reckoned with, and every reckoning had to contain violence. Thankfully, they did not look close. They did not inspect the creases and the folds and the in-between, furrowed places where the unlovely parts of healing hide. Where naked eyes can't reach, they did not, or would not, look.

For a long while, I avoided looking people in the eyes. I said very little that was true, even after I found one person on the planet with whom I could be truthful. To the others, I tried not to speak. If they said, "Hello, how are you?" I said nothing in return, just stood aside and held the door open for them with a cool, impenetrable chivalry. The rage stayed hidden in the furrows unless and until someone slipped in too close when I wasn't looking. Only then would I test them to prove they really loved me. When they inevitably failed my test, the rage would flare out like a lit plume of gas and singe all the hairs off their arms. When they left, the only thing remaining would be the already fading memory and the smell of newly burned skin. In this way, the rage protected me from the consequences of closeness lasting too long.

I never expected to be able to stay sober long. Hours bulged with craving spilled over into whole days, slipped into weeks, lengthened ever more into months then years of consecutive linear time spent trapped between my ears in an unaltered state. Surprises came in the ordinary, everyday routine that emerged from my sobriety like pavement hidden under a blanket of leaves, uncovered suddenly by a strong wind and a spring rain. Physiologically, I felt better than I could remember having ever felt when I slept each night and practiced basic personal hygiene regularly each day. After a lull for more than a year, I returned to writing with a ferocity fueled perhaps by the still hidden rage. Stories disguised as poems came first, rhythmic missives I read aloud at microphones in bars and coffee shops all around Chicago. I became a featured performer on the circuit and achieved a kind of status I stayed largely unaware of until publications and awards entered the scene. When I entered my first graduate program, I was unrecognizable (to someone who might have known me before) as a model student.

In the spaces where chemical emptiness once dominated, the compact buds of densely curled wings slowly grew. As time passed, and years expanded into more years, the tight buds opened painfully into huge, colorful appendages granting my launch into healing, a flight plagued by turbulence and frequent falls, but rendering me sky bound, nevertheless.

THE HOUSE OF ENDINGS

Remains

1.

Liquid fills my lungs, and I am getting too big for this space. I send the signal to her body that starts her labor. The signal is chemical or electrical or both. The signal is thoughtless, and it means I am ready to cross over from this world to that one. Ready to break open her pelvic bowl and exit this place. This place where I can't see her skin for being inside her skin. This place where I can't see. Swap fluid for air. Denser for lighter. Perhaps, when I'm dying, I will be in the womb of the next world and perhaps, when I'm ready to cross over into that world, I will send a signal. I suspect that is what happens, and I suspect it will involve letting go and lifting, lighter than air.

2.

Forty-three years later, dressed in gold Doc Martens, blue joggers, and a tank top, I walk into Hazel's apartment complex carrying my blue Cambridge Hospital Maternity Ward tote filled with crystals and a mala. My hospice volunteer ID badge is affixed to the waistband of my joggers, facing forward. I want to strike a balance between looking official and being myself, and it's *July-wanna-die* in San Diego. Every year, we have *May-grey* and it's cold, *June-gloom* still cold, then we're hit with *July-wanna-die* ultra-heat, reminding us, despite the ocean teeming and roiling alongside us, we live in a desert.

The door is open, and it is dark in the apartment. I lift my hand to knock, and get three knocks in before I hear, "Hello, Elizabeth, come on in."

The room is small. A skeletal woman lies in a hospital bed along the back wall; there's a large and loud television along the left wall, and couches stretch along the length of the other two walls. Seated on each couch are Hazel's daughters, Belinda and Chloe. They tell me their mother has Alzheimer's disease, and has been declining with dementia for years, but is now, they suspect, in her final weeks or days of life. Chloe is the oldest of five, herself a grandmother at age seventy-two, long and thin with short gray hair and impeccable posture. Belinda is the youngest, age forty-nine, also with short gray hair, also long and thin, and with a chiseled jaw line and expressive mouth that bends into shapes to announce whatever it is she's feeling. "I'm the youngest of five kids too," I say. Belinda pats her chest with a fist and winks. Both daughters look tired, dressed in sweatpants and grief. Hazel, the skeletal woman in the hospital bed, their mother, is sleeping. Her skin is wrapped so tightly around her skull, her cheeks so sunken and her lips so thin, I struggle to rest eyes on her without alarm registering on my face or body. I sit beside the bed feigning expert ease and feel the heat in my hands.

3.

"That's some tone-policing, white fragile bullshit!"

You are my eleventh love, and we are in our eleventh and final month of our romance. You yell these words at me with a face crumpled in rage. Or is it anguish? Your hair is flat against your head and braided in the back, accentuating the masculine in your face. Other times it's free and lifts away from your scalp in perfect helixes like rising smoke. You call it *a mess*, but I call it perfection. Your broad shoulders and muscled arms are unmoving, as though carved from marble, and you wait. Still, but for the small beating under the tender skin near your neck, the light rhythmic lift of the thicker skin on the back of your arms crossed over your belly. You wait for my response. I would have preferred it if you had slapped me, or even punched me. For a small terrible colossal moment, I think I might yell back, *Just hit me,* like I had half my lifetime ago when my heroin addict girlfriend put me up against a wall by the neck and cocked her fist, ready to fire. I had yelled that then because I'd felt a similar feeling in my chest, something hot and coiled that suddenly sprung and made me wild, crazy to be hurt in a way that didn't hurt quite as much. In a physical, acute way that comes with a cut or a bruise attached to it. But I don't yell *Just hit me.* I pull my spine into a taller, straighter line and struggle to suck the tears back into my eyes. Your face is expectant, daring. Daring me to say one more fucking thing and promising me that it, too, will be wrong.

"OK," I say. What I don't say is it isn't fair of you to shout those words at me in the context of our breakup while I'm crying and after I said *You're too hard on me*, because it was the most honest thing I could think to say when you asked why I was crying. What I don't say is it isn't fair your tears are vulnerable and brave while mine are fragile. What I don't point out is the double standard, which isn't the only double standard, but it's the most impossible. Because the heartbreak I feel over breaking up makes the crying happen, and it doesn't have anything to do with my race or my ego and the fragility of both. Although, I do know that when white people say that something isn't about race, it usually is actually about race if you look closely enough. But I don't say anything because I know what you would say. *Oh, it's not fucking fair, is it? Welcome to my universe!*

I want into your universe, or any universe other than this one. Any universe where we could have been treated the same. Then maybe we would have had a chance. But we are equals because when I stand and lean my body into yours, I feel the upright nature of a home with skin in place of walls. Maybe walls would have been better because then old paint would just flake off and fall to the floor like silence. You object to pronouns that create artificial divides, that don't apply to anyone, least of all you and also me.

We are like our ribcages: two mirrored sides connected by a spine.

We are like our spines: connecting the left and the right, while being neither and both.

4.

I am a child, and I think babies grow inside their mother's stomachs, in a sort of reverse-digestion process. Rather than a larger object being broken down into its component parts, converted into energy to animate the body, then the elimination of what remains, instead, what remains is a magical, seeded soil, fertile and life-making. Even now, as a child, I know this process is magic and women are goddesses. How else could anyone explain a whole new human being created out of what remains in a woman's stomach?

5.

There is so much we don't see. The human eye can only see visible light, but light comes in many other "colors" — radio, infrared, ultraviolet, X-ray, and gamma ray — that are invisible to the naked eye. On one end of the spectrum there is infrared light, which, while too red for humans to see, is all around us and even lifts off our bodies like steam from hot water.

Stars are colossal; many we see in the night sky are millions of miles in diameter. Even the nearest stars are more than 24 trillion miles away and are therefore so diminished in size our eye can barely resolve them.

Our eyes have approximately 126 million light-sensitive cells. Of all the possible photon wavelengths out there, our cone cells detect but a small sliver, typically in the range of about 380 to 720 nanometers — what we call the visible spectrum. Below our narrow perceptual band is the infrared and radio spectrum, with the latter's longer, less energetic wavelengths ranging from a millimeter to kilometers in length.

6.

My child logic takes me much farther, in fact. I wonder, what foods would I need to ingest and in what order to produce the right mix of elements for the alchemy? Could I accidentally make it happen in my own stomach? It isn't long before I learn babies are made in a completely distinct organ, and food has no part in conception. Somehow still, I eat two bites of an apple, a bite of a fruit roll-up, and a ketchup packet, timing it all perfectly and placing my hand over my stomach after the final swallow, at once hopeful and afraid. And when it doesn't happen, I worry it is because I am too empty. Too not enough.

7.

"Is it OK if I give her Reiki?" I ask.

"Oh yes, please, we would love that. They told us you do Reiki," Belinda says.

I place both hands on the leg closest to me and close my eyes. The leg feels so thin, so fragile, I worry I will injure her with the weight of my hands, so I hover them, just touching over the sheet. And with that slight effort of holding the weight of my hands and forearms aloft, that nearly effortless effort, I feel it. Rather, I sense it, a similar level of effort made by Hazel, but in reverse. Rather than holding up, she is holding it all down — tethered to the long, slight body, the remains of herself. As though by a string, as though she were a balloon filled with a lifting gas — a vapor density lighter than air, like hydrogen without oxygen. Like helium. Like nitrogen. Like water vapor. Like smoke. Like a ribbon of smoke, rising from her remains smooth as a slip of velvet then curling, coiling, spreading. Decorating the air. Floating around the room, dancing over our heads. She is such a presence up there; I almost want to lift my face to look for her.

"I'ma get ready for work," Belinda announces and walks out. She lives in another apartment in the same complex. So does Chloe. Half an hour later, Belinda is back wearing her Mass Transit Authority uniform. Her hair is now shoulder-length, straight, and black.

"It's a wig," she says and puckers her lips, pats the hair. "Been driving that bus twenty-seven years and ain't nobody never seen my real hair," she laughs. Chloe laughs, too. Their laughter fills the room for a few bright moments, and we're all lifted.

I glance at Hazel's face and her eyes crack open. Just then, she crashes back into her remains and opens her mouth. Beneath my feather-light hands, her leg tenses, and the rest of her body tenses, and her arms lift and, like an animated mummy, she rises, lifts up the top of her torso, reaches, eyes wide open now and fixed on the wall. Fixed on something in between her face and the wall. Fixed on something that should be anything other than the wall.

"Ah-tie," she calls out. "Ah-tie, ah-tie!"

8.

When I learn about the way a woman actually makes a baby, it seems even more magical than my reverse digestion theory. Two single cells combining and then rapidly dividing in such a scripted way as to meticulously build, piece by piece, this intricate biological system wrapped in skin? What makes the cells divide? I spend hours at the library paging through encyclopedias and anatomy texts trying to understand. I learn about cells generating electricity, shooting out ions, heat, and light.

9.

When I meet you, I come with a past and its framework, its four walls filled with women, everything I've learned rendered useless here. Swallowed by itself. Swallowed by us two, too, because we eat our memories with some cheese and salami over a new kind of conversation. And almost without noticing, we digest them. Take whatever nutrients they offer and use the rest as compost: fertile soil in which to grow more words, more conversations, especially nonverbal ones carried on by fiery bodies. At the climax of those conversations, I see the future jeweled and dazzling as we hold hands and walk, blinking into the light.

Right after I meet you, the world shuts down from a global pandemic, and then the world erupts with an outcry for racial justice after yet another unarmed Black man is murdered by police. On the news, we watch as cities and towns across the country declare racism a public health crisis. "Of course," you say, "racism is a public health crisis."

"That's true, but there are underlying conditions. Comorbidities," I say. You ask what I mean.

"Too many people in this country are stupid, which is a real problem," I say.

"Or maybe you're just elitist," you say.

"What I mean is, ignorance is a public health crisis. Thinking is a skill that needs to be taught. Teaching children how to think needs to become a core part of public primary education."

You lean back and stare at me. I grow anxious to comply, to please. I worry when you stare at me this way you are critiquing me, though I also don't want to miss what the content of those critical thoughts might be. But when you open your mouth, you say, "You are so good looking."

Though I'm pleased you're pleased, the round pleasure is flattened by the disappointment of how much caring about how I look matters. It's like if you wrote a book and handed it to me as a gift, and I took it in my hands, stared at it and said, "I just love the cover." It's the one part of the book you're not responsible for. You'd think, *What about the story? What about what I created inside of that cover?*

10.

What comes to mind is the parable about the goats and the tigers, perhaps from some branch of Buddhism, which I studied and practiced in my twenties and some of my thirties. In case you don't know it, I'll summarize: A pregnant tiger pounces on a group of goats with the intention of eating one, only in that moment, she labors and births her baby and does not survive the process. Her baby tiger is born among and grows up with the goats, believing she, too, is a goat. She eats only grass and as a result, is malnourished. She learns to graze and bleat like the goats. Until one day, another tiger comes to prey on the goats and although they all run for their lives, the tiger who thinks she's a goat stands her ground, curious. The grown tiger asks her why she is living as a goat. She doesn't understand, so the grown tiger takes her to a still pond to show her who she really is. The grown tiger then feeds the malnourished tiger some meat, and although she chokes on it at first, she consumes it and lets out her first tentative roar.

One implication is who we really are is hard to swallow, and inherently tragic, because life can only subsist on death. Another implication is we are all tigers attempting to live here as goats. The first problem is, how do you live with these goats? The answer is, by knowing they are all tigers, and living with that aspect of their nature. The answer also is, perhaps in your art, you can let them know they are tigers.

11.

You are non-binary and go by the pronouns, *they* and *them*. You have a mountainous intellect, but you can sometimes fall into a well of thoughtlessness.

"I'm joining an investment circle called the Sisters of Africa," you say one day. Your face is alight with excitement and your eyes gleam with the promise of free money.

"What's that?"

"It's a private group of women of color who help other women of color by pooling their money and giving it to one of the members. It's a way of sharing wealth."

"Is it a pyramid scheme?"

"No, it's hard to explain. But picture a physical circle of women with one in the middle. It's where each member of the circle gives, say, twenty dollars, to the woman in the middle, and she keeps all of that and then gives only twenty dollars to the next woman in the middle."

"OK, so it's a closed system with a finite amount of money simply shifting around the system from person to person?"

"No, it's hard to explain, but there's always more money."

"Then there must be new people being introduced, and then more new people, and on and on exponentially. That's why it's called a pyramid scheme. Because there are a few people at the start and it widens from there, and the only people making all the cash are the handful at the beginning of the ever-widening pyramid."

"No, you don't understand. Never mind."

"I don't want you to get screwed," I say.

"Just don't talk to me about it. I don't want negative energy around my investment circle."

I say nothing more. Not because I'm wrong, but because I'm desperate for your approval, which I realize in this moment is another blind spot for me. This experience of being desperate for approval from Black people is a common dynamic for white people. Still, I know I'll never acquire it. Because too many people who look like me — white — have behaved terribly to you, and this too is perhaps a blind spot. Because blaming other white people for my interpersonal situation can be my impulse to deflect personal responsibility. What remains is your simultaneous love for and hatred of how I look, which is the source of the unconscious chemical signal in you that started this romance. And it is the source of the unconscious chemical signal in you that will, together with my blind spots, kill it.

12.

"Ah-tie" she calls again. "Ah-tie, ah-tie!"

"It's ok, Mama, it's ok, lie back now, just rest. Rest, Mama."

Belinda gently presses her mother back down and strokes her face. Hazel's jaw works, as though she wants to say more, but sound won't come. Her eyes water and seem almost lucid. She looks at her daughter, then at me. Her brow wrinkles in confusion. Suspicion?

She settles back into the sleep where she lifts again from her body and fills the air. Two more of her daughters arrive, as well as two men — the pastor from her church and the pastor's son, a preacher in training. They eye me suspiciously, my tattoos and scars. They smile false smiles and the pastor says, as though daring me, "Let's hold hands and pray." We all kneel on the floor and hold hands. I take the hand of my fellow youngest of five, and the hand of her sister, the oldest of five, and I feel suspended between them like the middle three, suspended between the loss of their mother and the different kind of loss of my own, suspended between that time and this one. And the preacher's voice booms. There are words about fathers and sons, but the most powerful presence in this room is a mother and her daughters.

13.

I find a series of beautiful, colorful illustrations of a developing fetus. They seem to move and dance on the page, ignited by my imagination and reverence, and I see the tendrils form, nerve cells growing long, wire-like extensions — axons — for carrying electrical signals over long distances. They then pass signals on to other cells by releasing chemicals such as glutamate, but they do so where they meet them, at synapses. That means the chemicals only have to diffuse across a tiny gap, greatly speeding things up. And so, there in the womb, long before birth, the nervous system is born.

14.

I am a child, and my older sister is very hurt in an accident and has become permanently disabled. My mom falls into the hole of my sister's hurt and has become permanently lost to me. I tunnel around in my sister's hurt, and in my own, searching for my mom. But the tunnels are dark, and I can't see three inches in front of my face. I learn to move by feel, alone. I find many things. None of them are ever my mom.

15.

The praying is done and the men are gone and Hazel is once again saying "Ah-Tie," but mumbling it now, with her eyes fixed on me.

"This is Elizabeth, Mama, she's here to help you. She's a nurse," Belinda says. I flinch, wanting to open my mouth to correct her, tell her no, I'm not a nurse, but Belinda puts a hand on my arm, a gesture to silence me. This is what Hazel needs to hear.

Hazel calms down and her eyes flutter closed again. Belinda reclaims her position on one of the couches.

"It sounded like she was saying 'I'm tired'?"

"She was saying 'Aunt Taye.' That was her aunt who raised her, closer to her than her own mama. She's been reaching out and calling to her all the time lately."

16.

It has been theorized early multicellular, complex organisms consisted of a digestive tube with nerves and a neural network — the enteric nervous system. This was the first brain, in theory, and the second brain, connected to eyes and ears, evolved initially to help us better feed ourselves. To better sustain those digestive tubes we are.

17.

Also, I see as a strange blind spot that you refuse to get a flu shot because you did it once and got terribly ill and then read somewhere vaccines cause all kinds of health problems. A clinical scientist by trade, I spend about an hour explaining to you the way clinical trials work, and how a tremendous amount of safety and efficacy data must be collected before a medication is approved for commercial use. Vaccines are based on decades-old research and thousands of people who have proven them safe and efficacious.

"Have you heard of Tuskegee?" you ask, referencing the infamous (and horribly criminal) medical experiment that was conducted on six-hundred Black men in Alabama in the 1930s. Men who had Syphilis were recruited and told they would receive free medical treatment, but were left untreated, even after it was discovered that penicillin was an effective treatment. Some even said that some men were deliberately infected with Syphilis and then left untreated so that the progression of the illness could be studied. Eventually, they all died of the disease. It was the first documented mass medical murder of Black people at the hands of white people.

"Yes of course, it's mandatory learning in every Good Clinical Practice training I've taken, which is about twice per year. It was the beginning of a series of strict law reforms, including the Sunshine Act, which forbids anyone from carrying forward hidden agendas in clinical research."

"I don't care," you say.

I stare at you across the restaurant table. We are seated outside, comfortably away from the next nearest table, but a warm breeze carries the sound of our conversation to them. I see their heads turned in my periphery. I lower my volume.

"But viruses are real. Infections, they're real. The most important inventions in Western medical history are antibiotics and vaccines. It's mostly no longer necessary for humans to die of infections, and often, from viruses, because these life-saving medicines exist. They exist by way of publicly disclosed clinical research conducted with rigor and oversight."

"We just don't know," you say. "Let's agree to disagree."

"So, when a Covid vaccine is available, you won't get it?"

"Absolutely, no."

"But this virus is killing people. And even people who survive, some have long-term, chronic illness."

"We all have to die sometime."

"You're going to play Russian roulette with your life?"

I feel my face reddening and hear my voice getting louder. You glare at me, your face like a wall. Like a stop sign. Like an immovable mountain.

18.

But I have no righteous leg to stand on, because nine months earlier, I am imitating my own corpse on a chrome operating table having the fat removed from my midsection. In bold spite of my intellect, some part of me believes it will improve my chances of having and keeping you. But I don't know that. What I think instead is I am fixing my body to match the aesthetic of yours. You — my new millennial non-binary partner — had sent me a naked photo of yourself, and it aroused as well as scared me. The image on that screen was terrible in its perfection. You looked like a body double, the person who stands in for the main actress to do nude and sex scenes, exclusively to be naked.

I ask you whether you really look that way without clothes on, or is it the lighting and the filters. I tell you I hope to see some imperfections in real life, lest I feel too intimidated. You tell me perfect is a story, a game. You aren't here to play it, you say. But then, in the next breath, you tell me you think you will be the next Toni Morrison. Sometimes you are too small, other times, too big. Too grandiose.

Edges blur. The air turns satin. The Valium is taking effect. Sober people like me, who work a recovery program and commit to change who we are, not merely abstain — we call this a *freelapse.* The cosmetic surgery center is different from the hospital environment, which is the only other place I've ever been prepared for a surgery. There, I was never given a Valium. Here, it's like the appetizer given before a fine meal. The blurred softness gives way to the blank in-between, and I am kept alive by a tube forcing air in and out of my lungs. I see my body as though from above: already thin, naked, pale skin surrounded by mirror-clean metal while a tube is inserted under my skin and yanked around to tear apart and suction out the fat.

19.

Three times per week for five weeks I visit Hazel. Belinda and Chloe are always there, always welcoming. We talk about everything: their childhoods, how their mother worked for the public school district for forty-five years, starting as a cleaning lady and working her way up to the director of janitorial services by the time she retired. Belinda does most of the talking. When Chloe speaks, Belinda bites her lip, as though stopping herself from interrupting her older sister.

"I hate losing my mama," Belinda says, "but sometimes I hate seeing her like this almost as much, and I wish she'd just let go. She keeps holding and holding on."

"Have you told her it's OK to go? That you're ready and you'll be OK?"

Belinda's eyes fill and spill tears so fast, so immediately, her expression barely changes. Weeping, she bites her lips and shakes her head. We both look at Hazel.

"Aunt Taye. Did you know her? What was she like?"

"Oh yes, we spent summers with her. She was a midwife, and women came to her house to give birth."

"That makes sense! She's there, on the other side, midwifing her birth into that place."

I feel it in my gut, the tingling wonder I felt as a child in the library poring over books and illustrations about gestation. What is gestating there, in the place where Aunt Taye is already visible to Hazel, half in this world and half out? Is it chemical? Gaseous? Is it as vapor, in the air all around us?

20.

I have taken on a new client, a start-up in San Francisco doing groundbreaking research on neurodegenerative diseases like Alzheimer's and Parkinson's. One area of research is focused on the symbiotic health of the first and second brains — the gut and the cerebrum. The stomach, it turns out, performs many more functions beyond digestion. No, it doesn't grow babies, but it does contain a vast colony of 100,000 billion bacteria, the microbiome, which has a key role in the functioning of the enteric nervous system — the first brain. Experiments done with mice show certain types of bacteria in the gut influenced them — aggressive mice turn calm and calm mice turn aggressive, just based on the presence or absence of certain bacteria. Similarly, the degeneration of neural networks around the stomach and intestines is observed in certain mice when an imbalance is introduced to their microbiomes. Some brain diseases, like Parkinson's, may stem from the degeneration of intestinal neurons.

21.

That night, I stand before a mirror and see the effect: my brutalized torso. Its shape is the same, but smaller, and very hurt. The bruising is extensive and deep, echoing back the wails of deeper wounds still too fresh, trauma carried long in the body. Red and purple and blue and black, my skin throbs, both anew and in remembrance. The new wounding beckons the old, and not even that old, but merely two years before, when I was torn apart and fitted with metal and mettle and my skin, stapled and knitted back together, was similarly bruised everywhere. The trauma that still lives under the surface rises — the helplessness and fear and grief surface with it. I fall into my bed and curl up tight. "I'm sorry, I'm sorry, I'm so sorry," I say to my skin.

I am alone with my skin for three weeks. I take this time away from my anticipated lover with no real explanation, initially, but then I decide that I must tell them something. Perhaps that I'm recovering from some other type of outpatient surgery.

The surgeon has used the scar over my pubis as an entry point. At each end of it is a single stitch and a scab. I lay my palms over them and close my eyes. My body shines and shivers with hurt. My mind is nothing but an apology.

22.

I had been working out daily, mainly to rebuild the muscle that atrophied when I was in the hospital, so with the fat magically removed, what remains is muscle. What had been unfirm is firm. Abdominal muscles suddenly looked so prominent in the mirror, they cast shadows on my skin. To the touch, the place just under my skin is hard, not squishy like before.

What remains has lost its softness. Even the fat that remains is hard, and tunneled. There is a strange pattern of corrugations under the skin of my abdomen, which are from the suction wand creating small tunnels around which some of the fat hardened.

I am appropriately unhealed when I join my skin with yours for the first time.

The bruises are still there, and I lie about their origins. I tell you and others I had an outpatient procedure to remove scar tissue from around my pelvis, where a couple of years earlier, I received orthopedic surgery to rebuild destroyed structures after a major motorcycle accident. A small cage-like arrangement of titanium plates and pins replaces my pubic bone. A large screw re-connects the bones in my left sacral-iliac joint. It seems reasonable and even likely a subsequent, small orthoscopic surgery to remove scar tissue would be beneficial.

In a video call with you, I tell you a half-truth. "I'm saying it was to reduce pain, but it was really vanity that drove me. My pubic bone protrudes. I want to change that if I can before you see me naked."

"I don't care about that. You know that, right?"

I nod.

"Your *pubic* bone was crushed? What happened to your *vagina*? Did it live?"

I laugh. I nod. I consider telling you the story of how narrowly it escaped death. I choose to keep it in the realm of mystery, where it can stay more appealing, like a filter on a lens.

Can it be true you don't care about the appearance of the object of my body, and what it contains and what radiates out from it is what makes it appealing? Why is it OK for me to care vainly about my protruding pubic bone but not OK for me to care vainly about fat, which has likely lived there just under my skin since my childhood? No answers come, only more questions, as I uphold the dishonesty, and as I suspect I am lying foremost to myself.

23.

There was the near-death of my body as a whole coupled with the near-death of my vagina, which had been cut off completely from its nerve supply and left entirely numb. When my pubis was crushed, a sliver of bone sliced out from the inside of my groin, high up inside my vagina just under my cervix. It sliced through a major artery, and I almost bled out. It also sliced through a major nerve, which cut off all sensation to my genitals. A gynecological surgeon repaired these severed lifelines but told me afterward that my chances of recovering sensation between my legs were about 50/50.

My vagina was resurrected, along with all the pleasure and pain it sources. And when I meet a woman with a matching arm scar, who had fallen while rock climbing and crushed her forearm as well as broken her back, I tell her the story.

"I almost had to go on living with a dead vagina. Can you imagine?"

"I have no feeling in my vagina," she says, and my face burns with shame. "I'm so sorry," I say and put my hand on her leg. She looks at my hand on her muscled thigh like it's a strange bird that has landed there, a curiosity. I remove my hand.

"I don't care at all," she says. "I don't even remember what it feels like. I don't miss it."

Refusal of grief. Forgetting as a chosen blindness. Can't miss what you can't see.

24.

Ninety-five percent of our body's serotonin is produced in our gut. It's the gut that regulates our moods more than anything else. Just as our emotions influence our gut, our gut influences our emotions. The wastebasket diagnosis of IBS is now thought to be a communication problem in the gut-brain couple.

25.

I know the beauty of breaking open and bleeding then healing over where I bled. My scars remind me every day how healed I am. They blare it loudly, telling me to care more, to heal more, to never stop healing, to not go back to slumping into the artifice of the day-to-day. They are a new kind of loudness made from lithe, awful, shiny skin. And yet to you, they do not diminish my attractiveness. You tell me often how attractive you think I am, and I worry you are too focused on this one aspect. When you sit me down to go over your list of grievances, my appearance is not on the list.

26.

"How long you been doing this?"

"About a year," I answer. A half-truth. Because her mother, Hazel is my first hospice client, but the training program took about six months to complete, and because of Covid, another six months passed before I was assigned to a patient.

"And you ain't get paid no kinda way?"

"I volunteer."

Somehow, the holding of space and bearing witness at a labor and birth seems similar to that at a dying and death.

As I lay out crystals alongside and atop Hazel's skeletal frame, I talk about the science of Alzheimer's, the plaques and the tangles, and how the mechanisms of disease aren't yet fully understood.

27.

I am in a hot tub, which seems solid, and my body less solid, and the water, liquid, but there it goes, lifting and floating away from itself in droves. Burgeoning clouds of mist that had just been water, mist in such mass quantities I think the water might soon drain from around my porous skin. I lie back and float and join my body with the water and it, too, is liquid. And it, too, rises, floats up and away. And the stars above are present in an impossible quantity. More than the bright beads on the surface. Beyond that, another layer, more multitudinous but dimmer, and a larger group in a layer beyond that, so they seem to bulge into one another, expanding and contracting like lungs, like tides, and my eyes make them shimmer, twinkle, dance, even shoot clear across the black sky. And the mist is among them. And the water is among them, and the vapor, and I am, too.

28.

What if, therefore, the story or game we are playing out on this stage of space and time in these costumes of flesh and blood is worthwhile? Providence? Dense with meaning?

In other words, what if the great human experiment is counting on you and me to push the boundaries of what we can do with our roles? Roles we were born into. Roles assigned to us at birth. Roles that the tone and the mood of humanity cast us into.

Maybe I want you to care about the game and play it and, in doing so, push its boundaries and change it from the inside. Because, despite your blind spots, you are ten times smarter than the average person and you aren't afraid of hard work and you will win the game in bold spite of its deck being stacked in favor of whiteness and maleness.

29.

You say winning this game cannot bring you joy because it's not your game and its audience is not your audience and what you're doing your work for is so much deeper and more important. Maybe Toni felt the same way. Maybe that's why she didn't publish her first novel until she was thirty-eight. Maybe she had already said what she said in *The Bluest Eye* in her Master's thesis at Cornell when she highlighted how Virginia Woolf and William Faulkner, two of her greatest influences, treated "the alienated" in their work. Maybe, when she was Chloe Ardellia Wofford, experiencing colorism at Howard University, and subsequently studying English at Cornell, she started searching for the right words to put together in the right way to deliver the message perfectly. To deliver the truth through fiction. To change the world.

She wasn't writing for a white audience. Nor was she writing exclusively for a Black audience. She was writing, I believe, for the progenitor of the human experiment. Whatever forces created this relative world and its insistence on dichotomy. Whatever tore us apart from our source (each other?) and scattered us across a hostile planet with more velocity than we can counter and upon which we have no choice but to play. Our roles. The game. And yes, part of your participation is to feel the trauma and the rage of your mother and her mother and her mother before her and back and back and to carry that into your writing where it can point to something transcendent and make that transparent for the average intellect. Break open a crack on the surface of what appears to be so the light of what *is* shines through and blows their minds. Change them permanently and forever.

It could be that something along these lines is what's behind our mutual drive to write. To, as Toni says, *do language*. Because we can feel its potential, its potency, its inherent power. The ultimate truth, she knew, is going past what we've been taught. It's a simple concept, yet we are so used to being taught something else the words tend to block truth rather than letting it through.

30.

The other problem is the sorrowful game we are here to play in time and space. If we try to correct the sorrows, we only shift them elsewhere. What we are left with is the best we can do: to recognize the eternal within ourselves and others and participate with joy in the sorrows of the world. Like in the parable, we are tigers staying alive by way of death. We play the game. It hurts, but we take it less seriously, because we have found the transcendent place inside.

31.

As I offer this parable — this strategy and invitation — there is another part of me judging it as a form of spiritual bypassing. And the only way I am afforded the ability to bypass such problems with spiritual axioms and parables is the privilege I received when I was born into this skin. And like you say, it doesn't matter how intelligent, eloquent, polite, composed, talented, or beautiful you are in certain situations when you are Black or Brown — the people with your life in their hands won't see any of that. They will see only their programming. And whether you see your true eternal self and theirs in that moment or not, it's not up to you whether you can stay in time and space and play the game. And so you opt out. Or, at least, you opt not to give any shits about the outcome of a game so unfairly rigged against you for so arbitrary and absurd a reason and you refuse to extract any joy from its fruits.

What I want to know is, what happened to that mother tiger who died? And what filled that space in the baby tiger? Or did she get stuck there, in that hole? Waiting for something to fill it. Sad and frustrated that nothing ever did.

32.

You tell me I am fetishizing Blackness because I'm attracted to you.

Fetishizing Blackness?

Or is it a biological imperative to be attracted to people who look different from me? To diversify the gene pool? The more homogenous the gene pool, the weaker the offspring. Are we driven by nature to couple with dissimilarity? And are we programmed by culture to couple with resemblance? Even as I offer this rationale for my attraction to you, I hear how it comes off as dismissive and not listening to your reality.

My instinct to turn to science as a defense when someone says they feel fetishized is an intellectual failure on my part — another blind spot.

"There's what we can see," you say, "and then there's everything else."

But I think it's the opposite. There's what we cannot see, and then there is the small opening of what remains.

33.

The implications of this research are so far-reaching, it's staggering. Think of how long human animals have been on the planet, even in primitive, unevolved form, and the genetic evolution our DNA or genome has undergone. About six-million years is the theory. It's like an aging wine: the older it is, the more robust and complex. The human genome, likewise, is much stronger and wiser than it was at its inception, having learned from its changing environment over time and selected adaptations to make it better at survival. We adapt to survive and survive because we have adapted.

34.

"How you know all that?"

"I'm a scientist. I work in clinical research. One of my therapeutic areas is neurological and central nervous system disorders."

"And you also up in here with Reiki and crystals?"

I smile. "Yes."

"How's that work?" They laugh.

"There is vastly more that we don't understand about biological systems than the small fraction of what we do. And fundamentally, we know, everything is energy. Energy can be manipulated. All I'm doing here is putting the concentrated energy of love, like an electrical current through a wire, into her body through my hands. And certain crystals are thought to enhance or direct it in a specific way."

Both women nod.

"I believe that," Belinda says. "I believe it."

35.

Bacteria are a separate, way older organism. Who we are is not determined solely by our choices, the sum total of our actions and non-actions over the course of a lifetime, combined with our hard-wiring, our nature. We're influenced by the bacteria that were present on this earth for billions of years before us, that have acquired a genetic repertoire that's vastly greater than ours. And they, too, influence who and how we are.

36.

I want you to incorporate the inevitability of sorrow into your calculations and still be willing to play the game. To stand for who you are in every moment. Maybe the game is rigged, and maybe it's not fair of me to ask you to play it, but I am asking anyway. When Toni won the Nobel Prize for literature, she invited everyone dear to her to the ceremony. She owned the right to be who she really was.

The vastness of you is something I feel when I'm close to you. It feels similar to standing on the edge of a cliff that drops and opens into a bottomless canyon. The rush of adrenaline is not so much from fear of falling in as eager anticipation of that, and trust when I do, I'll be rewarded.

37.

I like to joke with you sometimes when you're in the kitchen with no shoes on, cooking for my kids and me. "I like you barefoot in the kitchen," I say. "Now all I need is to knock you up and you'll be just right."

You throw me a warning look, your mouth in a frown but your eyes smiling, yet with a knifepoint aimed at me still, because we both know your masculinity lacks that toxicity, but neither of us can claim, not with any confidence, that mine does.

38.

"I don't want her to be in a bag when they take her out. I don't want no sheet up over her head neither. Can you please just have her looking like she's asleep, and going to the hospital or something?"

I call all of the hospice personnel I have access to right then and there and convey Belinda's wishes about her mother's remains. When I'm done, I smile at Belinda.

"Girl, you a giver," she says. "Mama was like that. That's why she called you to us. You in, girl. You family now."

"Now that I'm family, I need to tell you something."

Expectant faces.

39.

Sitting in the chair at the ophthalmologist in front of the contraption for testing various functions of my eyeballs, including pressures and the thickness of my corneas, it occurs to me: this sense is perhaps the most imprecise. Both my pressures and my thicknesses were labelled "good," which made them feel optimistic I wasn't going blind. But what about that half-truth? The one I gave to my partner? Indirectly, I presented myself to them as a "woke" white person. One who understands racism and white supremacy and actively works against it. But I was only halfway through reading *How to be an Antiracist* when I met them, and I still had blind spots. And the remaining half of the book was the most important.

Blind spots.

This colloquialism is code for, *there are things you don't see.* Spectrums of light? Other dimensions? Neither and both.

41.

"I'm gay," I say.

"Yeah, we knew that. We don't care," Chloe says.

"It's obvious, girl," Belinda says.

I smile. "I wasn't sure if you would be OK with it, because I'm sure your religion doesn't see it as OK," I say.

"You know, God wants us to hate the sin and love the sinner. And you know what? I have gay friends," Belinda says.

"No, Belinda, don't say that. That's like when a white person says they have Black friends, don't say that shit," Chloe says.

I laugh.

"Oh, OK well. I mean, not to say that, but I have a niece? She's a young woman and she's always been a little different, you know? Now, she announced she and her partner are getting married, and she invited me to the wedding. She's the man in the relationship? So she's wearing a tux and her wife is wearing a dress, you know?"

42.

We are against the wall in my bedroom having shed most of our clothes. They lay in a heap at our feet. I lean my skin against yours and kiss you. The heat between us grows and we press harder together, come apart, press together, come apart. You make small noises while we kiss, and when I leave your mouth to kiss your neck, you whisper, "I'll be your sex slave."

I freeze. I lean back and look at you. Is this a test? A lump rises in my throat and I worry I'll be sick. I bend down to the floor to pick up my shirt and put it back on and you grab my forearm a little too hard. "Don't, please," you say and squeeze harder. "Just fuck me. I *need* you to fuck me."

The damaged nerves in my arm whiz and crackle, trying in vain to generate a pain signal to tell my brain, *Say no. This is bad.*

43.

If I press down on my closed eye with slight pressure, I can see my optic nerve. It's a glowing white circle against total blackness, surrounded by impressions of the leaf-like network of smaller nerves.

I'm pressing a tissue over my closed left eye to absorb the liquids and gels they've squirted into it, and this is what I notice. Push a little and I see outer space. Push a little more, and there's the perfect circle of burning white. Completely there. Completely in focus. What fascinates me most is the seeing of it. My eyes are closed. What's doing the looking?

"You have big cups," the ophthalmologist says. Assuming she's talking about something eye-related and not my bust, I ask, "Big cups?"

"Your optic nerves. The normal size is a three, see?" She shows me a diagram of an optic nerve. I stare at it trying to see anything other than a blurred circle. With dilated pupils, I realize I can't see well enough to read or make out hard lines and boundaries. They fuzz and blur and turn soft under my compromised gaze. Trying to focus only makes my head hurt. She flips the page to reveal a photo of a larger circle. "This is yours. It's a six." My mouth drops open. That's twice the size of normal. Unsure whether this is good or bad news, my expression settles around neutral territory. Not smiling, not frowning. Expectant.

"It's not bad, it just means you're at a higher risk for glaucoma. Do you have any family history of glaucoma?" The image that comes to mind is of clouded, milky eyes. I think of my paternal grandfather who went blind in his old age. It wasn't glaucoma though; it was macular degeneration. "No family history," I say.

This half-truth borne of fear reminds me of the time I lied to you about the motives for my elective surgery, and the origins of my vanity. I was willing to admit I wanted to cast myself in the role of body double, turning to angle my object in the light so you would see me how I saw you, but unwilling to admit it was fat I had removed, not scar tissue. The subtext was the same, I reasoned. Was I looking for attention? To be desired? Did I even exist if I wasn't attended to and desired? Did I exist if someone wasn't seeing me, and if I wasn't in control of that seeing?

44.

Maybe it's what we *don't* know that keeps us here, tethered as we are by gravity, magnetism, electricity, blood, hydrogen, oxygen, nitrogen — gaseous at the same time as solid and liquid, a dynamic zapping collection of deeply intelligent molecules holding secret codes, codes that while keeping us alive, are never to be revealed to anything living.

45.

One way unjust caste systems protect themselves in perpetuity is to selectively admit roughly one percent of the lower caste members into their echelon. These people then serve as mascots the zealots can point to and cite as proof these power structures that disenfranchise ninety-nine percent of the people don't exist. Now you see us, now you don't.

Or, they say, OK, yes, racism existed in the past, but it doesn't anymore. We are a post-racial people, holding hands across America. It's some bullshit.

46.

We have broken up, and you regret it. You text me and ask me for another chance. You tell me you aren't perfect, and you see how you were too hard on me, and you can do better. And although I miss you sometimes, mostly, I feel a great deal of relief. The kind one might feel, after banging one's head against a wall repeatedly, by stopping. Interrogating my relief, which itself also makes me sad, I wonder whether it's partially due to a retreat into white privilege and away from facing racism on a regular basis? I'm not perfect, either. I still have blind spots along with my tunnels, my self-made caves of hard fat ignorance to fall into, to get lost in.

Weeks after you're gone, you text: *Tell me something kind.*

I miss laughing with you, I reply. *We laughed a lot.*

You text back: *I miss your stories. And your abdomen.*

47.

One human is made up of thousands of genes, billions of neurons, and hundreds of billions of bacteria, all with biological continuity. We are trillions of tiny, dynamic links and bits of information snapping and zapping in perpetual motion at unfathomable speeds. This amounts to an overwhelming complexity that far evades the capacity of human understanding.

48.

Axons and neurons, fibers and proteins, plaques and tangles. Electricity. Energy. What is that? And what is left of us when all we have left is that?

What if life is nothing more than a stack of moments one on top of the next?

What if death is, too?

And what if all we can't see, can't know — all the undigested truth — connects these moments together like a string of beads, like a mala, like a chanted prayer?

49.

Hazel. Her remains. Her skin wrapped around her bones, all of it already breaking down, and how much does the shapes it makes matter? It is this stopping place of a body where the absurdity of vanity is thrown into sharp relief.

Hazel. It is not about her shape, her vanity. It never has been.

It's the software of her.

The raw code.

There it is in her girls. Belinda's bent smile. Chloe's hard stance. The other three daughters and their spiral staircases of sameness.

The day before Hazel dies, I bring five small heart-shaped rocks and charge them on her heart, pressing them to the skin on her ribbed chest, or what remains of it, allowing the energy of her heart to download and hold inside each rock, one for each daughter.

When she dies, they do not carry her remains out in a bag. They carry her out the way Belinda wants, in the hospital bed with a sheet pulled halfway up, Hazel looking as though she is sleeping. In all her forty-nine years, Belinda has never seen her mother look so at peace.

Hazel's expiring carbon remains in the room, on the carpet, in the air. Tethered no more.

Vapors.

Nitrogen.

Like a placenta.

Gathering ambient strength.

Gavage

Inside the story of my impetuous and regrettable decision to undergo liposuction to remove fat from my middle, lies the layered and largely unconscious motivations for such a choice. That my children will one day learn about this mistake obliges me to interrogate those motivations closely.

This is a story about gavage, the process by which ducks and geese are force-fed up to 2.2 pounds of grain and fat through a pipe rammed down their throats into their bellies. This is done multiple times per day, spaced out so that the birds will be over-stuffed but not enough to split open their stomachs and kill them. Instead, the birds will survive with a swollen liver. Livers that swell up to ten times their normal size, so big and heavy that the birds can barely stand in the end. Living in tight cages and close quarters, they're known to tear out their own feathers and sometimes attack each other from the stress.

Every human animal, no matter where in the world, whether born with a penis or a vagina or both, is made to drink misogyny from a fire hose from birth.

This is a story about force-feeding.

Whether overt or thinly veiled, dislike of, contempt for, or ingrained prejudice against women wove its way into the DNA of our ancestors. This, even though it all starts with women. Women are the creatures that make penises. I made one. My mother made several. Her mother also made one. And this power — the power to create whole people, whole worlds, entire lifetimes with glory and pain and heartbreak and redemption, with years and decades and whole afternoons filled with stories, with or without penises, all inside one's body — is threatening.

This is a story about what happens when people feel threatened.

The swollen livers that make the expensive delicacy known as foie gras can come only from male ducks. This is because males put on more weight than females, and the fatter the better for the buttery fatty spread for bread or fine crackers. The industry, therefore, having no use for the female ducklings, tosses them alive into a meat grinder to be used for fertilizer or cat food.

It's not that my motives for the mistake I made relate to the story about gavage, but then again, they do. It's in the brutality built into the fabric of humanity. It's in the way patriarchy and its byproduct — the unconscious or conscious hatred of women — harms males and females, alike. It's in the way life is sustained by death. It's about how much blithe exploitation and the willing infliction of suffering is involved in the production of food, and the production of beauty. It's about the disordered relationship so many of us have with both. It's about our compulsion to fall in line and deform and violate our own bodies.

Where the Truth Hides

Between the shadow and the soul, truth hides in secret. These are some of the words inked across your skin, my eleventh love.

Perhaps this is the place you haven't looked, where unconscious motivations for the hardness live.

Remember that night on the beach during our long courtship disguised as a friendship? I asked you to hold your palms up to mine and I looked into your eyes and in that moment, in that bright darkness, I peeked behind the curtain and I saw you. What I saw, I will never forget. And from seeing it, I will never be the same.

It wasn't your didactic monologues cataloguing my faults that changed me. It wasn't even your kindnesses, your varied creations made from words and wire and stone and thread you gifted to me. It was that glimpse of you that night on the beach — beside our shared mistress our shared womb our colossal other woman, the Pacific — with flashes of her tumult reflected in your face and the sound of her fury in our ears. A place I longed to get back to after, during our slow and certain undoing. The bowl of sad memories you were about to curl up in had a lot to do with the enemy you made of me.

Maybe it started that day Jacob Blake was shot seven times in the back at close range by a white police officer. You called me crying. Could I come over? On the way, I hit terrible traffic. The alert popped up on my phone and read: "Serious accident reported." I heard sirens. The car was at a complete stop, the highway had become a parking lot. I looked to my left and right and saw people with irritated faces, annoyed they would be delayed getting wherever they were going. *Whoever was involved in that serious accident, their life will never be the same.*

I had that thought, then let it go. No more thoughts of its kind, no memories of my accident followed. Instead, I turned up the volume on whatever audiobook was playing. The car inched forward, and I reached the place where the accident had been. A large swath of pavement was stained in blood.

The car picked up speed then as the traffic opened up, and the strangest thing happened. My body turned on a tiny faucet under every pore and sweat surged out of me so rapidly that within minutes, my shirt and pants were soaked. There was a puddle on the leather seat beneath me. I opened all the windows and tried to cool down, but I wasn't actually hot — it was a cold sweat. Impatient motorists whizzed past at high speeds on either side of me. I felt nausea coming on. I pulled off the highway and stepped out of the car, bending forward, hands on knees, preparing to vomit. Nothing came, though. I got back on the highway and powered through the sick feeling, but my eyelids began to droop. My body was attempting to abruptly shut down, not caring that I was operating a moving vehicle.

Eventually, I made it to your apartment. You were outside waiting with the parking pass, clearly annoyed at my lateness. I stepped out of the car and felt wobbly on my legs as I walked with you into your building. Relieved not to be driving any longer, I pretended everything was fine and went inside and held you while you cried and listened while you raged. After some time, the nausea returned, and I had to excuse myself to the bathroom, where I vomited so forcefully, it seemed all my internal organs came out my mouth into the toilet. You questioned me suspiciously. Had I been relapsing? No, I explained, and told you about the serious accident, the blood-stained pavement. And as the words formed and exited my mouth, the new kind of purely physiological panic attack I'd been stuck in gave way to the familiar mental kind, and I once again found myself hyperventilating and shaking and crying, this time curled on your bedroom floor, images of blood and splintered bone and sounds of sirens and the pressing close of death overtaking me. You said something along the lines of not having space for this, and walked out, closing the door behind you. Eventually, I self-soothed back to the surface where I could take measured breaths and slow my heart rate to normal and believe I was safe. And we sat in separate rooms, marooned by whole worlds and hundreds of years, each of us unable to feel safe for vastly different reasons.

Perhaps you were right when you suggested we should start over from the beginning. Perhaps we could have. Because a part of me continued to stubbornly believe we did not come together by accident, we had work to do here, we had something important for each other, something that would heal the wounds white supremacy and patriarchy had gouged into both our psyches. Although too, the wounds of white supremacy look different for each of us, and perhaps yours are better served by walking away. Or, as you say, back around to the beginning. Because if we don't complete that work in this life, we will likely get another chance in the next. Or in a parallel. Perhaps there are other, better versions of us taking that chance right now in some other world.

It's inside your tall intellect, like a rumpled mountain in the middle-distance casting shadows on itself, where you hide. A thing of beauty made more beautiful by its dance with light and dark, its sinew and contrast. Its weight.

We were formed from the same stardust, the same elements — your body of water and carbon matches mine. The metaphysical furnace that smelt you was different, but your heat under my hands was the same ninety-eight degrees. And the fire we found together burned us both the same.

You think all life is sentient but mine? Plants and dogs can get the best you have to offer because they lack things like words and much volition, making them controllable. I was uncontrollable with you.

If you had met me even a year earlier, I would have forfeited my autonomy, yielding to the power struggle our connection wound itself around. But I met you healed just enough to hold my own against it, the gravity of a mountain in the middle distance. Even still, my inner old-fashioned husband went behind my back to find ways to provide for you. My angry adolescent wanting to fight — when she sniffed out the familiar game, when she recognized the worn chess move of withholding — jumped out at you. These were mistakes I copped to even while performing them, which is progress. The only mistake you copped to was the ending of us with one short text. Later: I made a terrible mistake, you said. I'm sorry, you said. But the mistakes didn't start or end there.

I don't experience jealousy, you said. I don't understand it, you said. Love is not a finite resource, you said. But I watched the numbers and calculated a different truth, a hidden truth, a more human truth. It took a lot of math to do that — first formulating the equation, then solving it. All you needed to do was surrender your poker face and show your hand.

That last conversation we had in person, it took me by surprise. It wasn't what I expected it would be. I thought you would say, *I was wrong. I actually do feel jealousy. And the way I love you makes me so vulnerable, I'm terrified right now.* Instead, you invented ribbons of wrongdoing, again, and pinned them to my shirt, again. But there was scarcely room for more, and I was tired.

What I would have been able to live with was your hydrogen, your oxygen, the carbon molecules in your skin mixing with mine like the equals they actually are. What I would have been able to live with is your admitted jealousy. Because I would have related to that, and because I love myself enough, despite your chronically uncharitable assessment to the contrary.

And though I take it all personally, and yet have so much to learn, I wonder if at least part of the reason you were so hard on me is because you're hard on yourself, there in secret, between the shadow and the soul.

THE HOUSE OF BEGINNINGS

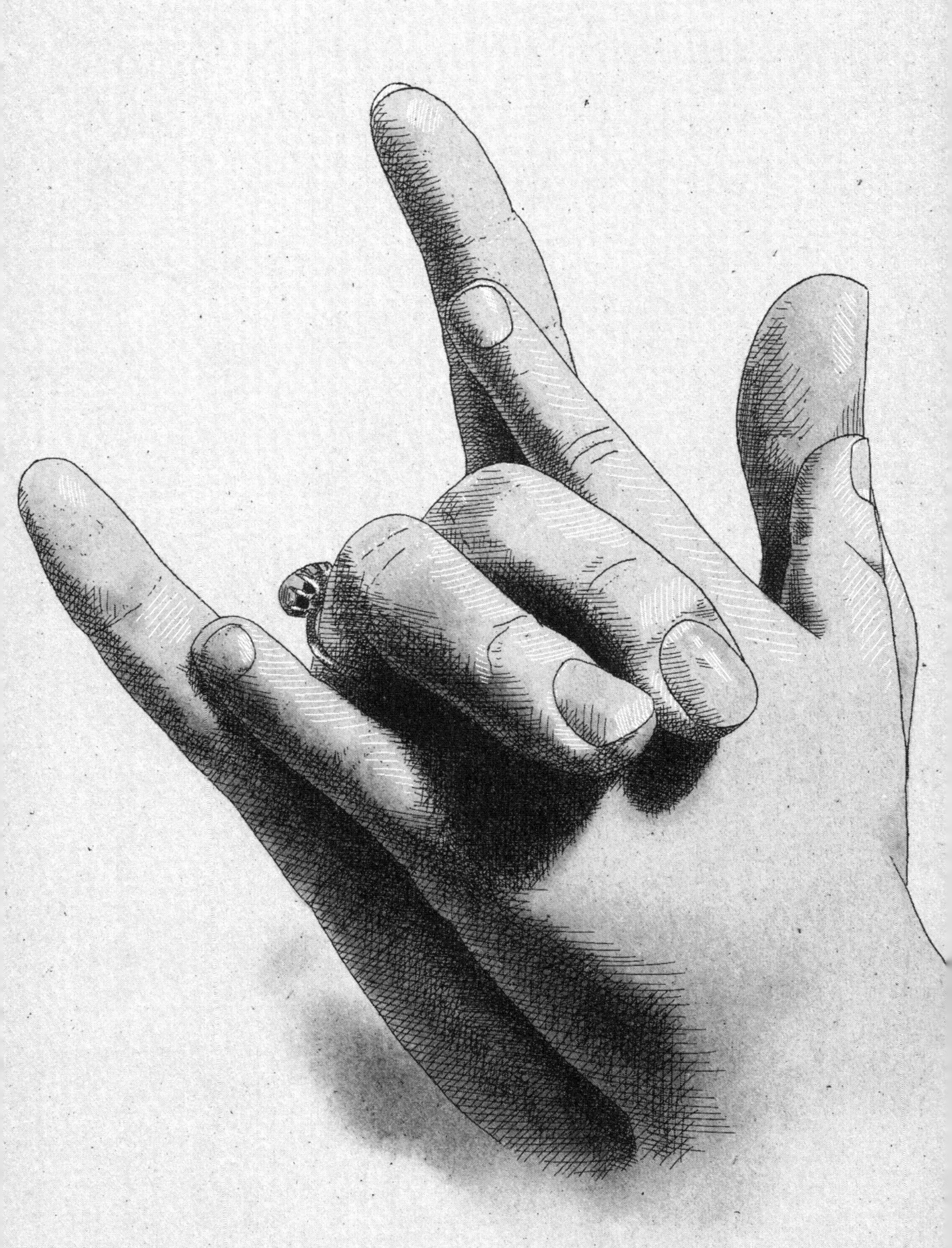

Give Yourself to Your Hands

1.

She is calm, my dog, and a champion sleeper. Her name is Snow. She sleeps like a koala bear, roughly twenty hours per day, depending on how much activity I make her do. She is always game for whatever I want to do: run four miles, hike seven, and she does get the puppy crazies and the zoomies sometimes when she's very happy, like at the dog beach. But her preferences are clear: her first favorite activity in life is to eat, second is to sleep, and third is to snuggle and kiss. She follows me from room to room, black and shaggy with loose tresses, like a curled shadow, her bright soul clearing a path for us, her zen energy the perfect counterbalance to my characteristic low-level anxiety. Each place I pause, she lies all the way down, relaxing every part of her compact body and closing her eyes. When I move on, she opens her eyes, unfolds, and stands, following me to the next place where she can lie again at my feet. She is the deep center of the nest I have built. We walk into the house after a run and I feel the energy nested there, warm. The dog laps water and I lunge for a towel to dry her beard lest it dribble all over the wood floor. The floor already shows the erosion of life happening all over it. Life in linear time.

I seek out my kids. My oldest is in a therapy session on Zoom; I know because of the white noise machine just outside their door humming its universal voice, its continual calming Om. In my son's room, I hear the buzz of the toothbrush, and I find them in his bathroom, his dad cleaning his teeth. My son's back is to me, his light-brown curls puffing up from the top of his head, face tilted up. His dad — tall and thin with short black hair and glasses — says hi. My son says a muffled hi, and in both laconic greetings I hear it, my nest energy. My warm.

2.

Hazel's funeral happens in a proper mortuary and funeral home. It is the family mortuary, where all the ancestors' remains have been handled, and lives lived honored. The sanctuary is packed full of bodies and the open casket at the front surrounded by bouquets of flowers is open. The red and pink and yellow and orange of the flowers seem extra vivid under the stage lighting, as though the light floral smell hanging in the air is an extension of color. From the back, I can see the remains within, the heavy makeup that attempts to reproduce the colors vitality makes in a body. It seems they have also done something to the remains to make the face plumper, which is disturbing. And on the heels of that judgement comes surrender, acceptance, understanding, because this is simply the custom of this place. This is how the dead here are honored. Like in some places where corpses are mummified and taken out each year, dressed up in clothes, and fitted with pipes, eyeglasses, mugs of steaming coffee. Whatever small things that person had done while alive, the surviving family members will reproduce with the mummy like a puppet being prepared for a puppet show. Then, they will stage a family portrait and take photos with the mummy. All the bright, round, smiling faces with the dry, withered, drawn face of the mummy smoking a pipe and wearing a fedora.

3.

Before the service, boys in suits pass out handheld fans and tissues. As the music starts, the family walks in, a procession of nine men followed by Hazel's five daughters and another woman. The preacher begins his sermon and explains this procession was made up of Hazel's nine grandsons, one granddaughter, and five daughters. "God only wanted good soldiers on his battlefield, and this week, he got one."

Then music, singing, weeping, standing, arms up, palms heavenward, the Father the Son and the Holy Ghost, yessir, yes God, "Therefore, be it resolved that ..."

4.

My new hospice client is Charles — my paternal grandfather's name. I walk into the small house with a sign on the front that reads "comfort care home" in handwritten purple letters. The house is cluttered, carpeted, furnished cheaply with blankets over the stained and torn couches. Charles is in what looks like the office at the back of the house in a hospital bed. His very large body is naked with a diaper that won't go all the way up. They had just given him a catheter. "He was peeing all over the room," his wife said. He doesn't understand the catheter is in, and thinks he has to pee. He is struggling to sit up while his wife and her friend are struggling to push him down. "Right before you came, he stood up and tried to pee in this trash can," his wife's friend said. I help them to restrain him and turn him onto his side because his face is contorted in pain, and we are searching for a more comfortable position for his body. He has a central line in his neck with a painkiller attached, similar to the one I had in the trauma ward after my motorcycle accident.

"Give him a bump," his wife says, and her friend presses a button. He instantly closes his eyes, relaxes. Then his wife gives him a pill, which she says is for his anxiety. "One every four hours," she tells me. She tells him I'm here, I'm a visitor, and he grunts. "He wants to be left alone," she says. I am made to sit outside his room in a chair. Charles is given a bell to ring in case he needs anything. The wife doesn't want me to give him Reiki, not even from a distance, and she doesn't want it for herself either. She's large, covered in dark sunspots, and has a lot of loose skin cascading from her arms and legs in pleats. She's in a bathing suit dress headed for their inflatable swimming pool in the back yard filled with water from a hose. She excuses herself to go and take a nap in the pool.

She walks out, and I look around. There are many books about Jesus and Christianity along with some paperback mystery thrillers. After a few minutes, the wife walks back in dripping in water and bursts into sobs standing in front of me. "I don't know what I'll do without him," she wails. I feel frozen in my seat for a moment and when I move to stand with the intention of offering her a hug, or a reassuring hand on a shoulder, she backs away and says, "You're welcome to take a plant from the driveway before you go," and disappears again into the backyard.

5.

Hazel's granddaughter delivers a moving eulogy, which includes a recording of Hazel's voice saying: "All you have to do is pray, and before you know it, God starts answering your prayers. You can count on it." Other than "Ah-Tie," it's the first words I've heard her speak, and the living voice crackling through the air in the sanctuary, coming from her previously alive body — it raises the small hairs on my neck.

6.

Charles is already gone. On my second visit, I bring my now-customary condolence gift for families: an edible arrangement of cut fruit in the shape of a floral bouquet. We sit in the living room on the couches covered in blankets and talk. His wife tells me stories about him. He was fun. He would pay the children to do silly things, like put their faces in pie. He used the money he got from redeemed recyclables to buy her flowers and he took them to her at her place of work. We discuss the possibility of whether there's more for Charles after his death, and I tell them I believe there is. We don't stop where our bodies stop. The wife agrees, but the granddaughter who works as a physical therapist doesn't. "I believe we get recycled, our bodies decay and become part of the earth and feed the soil that grows new life."

In this way, the body goes on, the body gets recycled, but I suspect there is something more for our souls. For our consciousness. Another place. The wife agrees; she says she knows it's true because she can remember a time before her present life. A place where she says, "they told me I would forget, but I didn't forget. I remember."

7.

My dog is curled at my feet while I work and I am in the warmth of my nest, the one I built independent of my romantic life. Yes, it did come together in this form while I was still with my ex-wife, the other mom of my oldest child, but it held together after she and I came apart, and now I have two gay men who are married to each other as my live-in co-parents and my two children and my dog. I get to have this nest and raise these children and enjoy having the help of two amazing co-parents to do the very difficult job of parenting.

After a second date with the woman from Tinder, I don't like her. She knows about DMT because she does a lot of drugs, and on the day we are supposed to have our third date, she tells me she's had three drinks and wants to see me naked. Although I haven't had sex in a while and I really want to, I decline her passive invitation. This body of mine, it's not for just anybody. It's strong and vastly more intelligent than I will ever be and imperfectly healed, but healed, nonetheless. I will care for it and cherish it for the rest of my time in it, and part of what that means is only allowing respectful, available, loving hands on my skin. Whether I hold this intention remains to be lived.

My dog will be curled at my feet either way, and she will love me the same, whether I wait for the quality of sex I desire or not. As though reading my mind, she lifts her head and gazes at me, all the depths of oceans there in her unconditional eyes.

8.

My new hospice client is a very large German man named Ewald; his wife is much younger than he is, a very small Mexican woman named Maria. The apartment is dark and small and filled, overfilled, bulging with this hulk of a man too big even for the oversized hospital bed in the center of one of two rooms in the apartment. He is groaning in a deep, gravelly voice. At his side, I feel my hands heat up and scan the mass of him for a spot. I settle on his right forearm. Maria thanks me profusely for coming and tells me she will go out for a walk and have a break. She will be back in two hours. When she leaves, he is half asleep, groaning every so often and taking breaks from breathing. When he stops breathing, I start counting, and sometimes I make it to ten or eleven before he draws in a ragged gulp of air and groans it back out again. This means he is close. I wonder if he will die under my hands, and I hope for it, as I feel it would be a great honor. I speak to his soul and say, *Fly free, Ewald. Your work here is done. You have done well.* He stops breathing again and I watch his chest expectantly. Again, the ragged inhale and gasping exhale. I begin to chant, and the dark room fills with the bright sounds of dying breaths and mantra.

9.

A new kind of meditation I'm trying is simply presence. What I'm present to: My life, and the sum of it, are immeasurable things. An open secret, a dream constant and common, it is precious, something one of a kind. No one other is exactly the same. I am present with the hush of my soul to find herself made now of skin and light, dressed in clothes, filling the weight of footsteps. I feel it in a strong way today, this strange and painful goodness of being alive. There is silence in the wrong places today. Places I am used to hearing a constant static are quiet. Something seems slightly different in everything. A crumpled piece of paper. My hands. The spine of this book. The sound of the wind. I have re-assembled just enough shattered glass to craft a sparkling new mosaic skin. There is a center within me that can hold steady all these imbalanced ghosts. I feel it when I hear the distant call of the train outside my window — the sound of determined motion, of going somewhere. I'm present with fear. Fear of death is really fear of love is longing for sex is longing for death is overcoming fear of love and of death because every living thing dies and comes apart. I came apart. And then I was put back together again.

My kintsugi, my repaired lacquer and gold seams where the cracks were — they're not just physical scars on my surface. They are joined fragments, more refined. Like a kintsugi pot, asymmetrically and perfectly threaded with gold, I am whole.

And, I feel so fragile so much of the time. I'm aware of how precarious all this holding together actually is. I'm aware, in time, everything will come apart again and fall away. And despite my getting closer to death, despite my little intimacies all in a row with dying people, I remain viscerally afraid. But this fear is not to be feared. Rather, invited in, sat with, served tea. Because when I hold it, and any other guest in this house of being human, I see how they are mere messengers. Teachers. Signposts. All conspiring to support me and keep me alive and thriving long beyond any reasonable expectation. And then I feel my power. Again, I feel how strong I've been and still am and will be. And this is the rhythm. This is the dance.

I'm available for it.

10.

There are people who say they can speak to the dead. They claim to communicate with them through something like telepathy, and claim the dead still exist beyond death as energy in some other realm. I am speaking with one of several I have paid to consult with, and fully expect her to be as fake as the others. Her opening sequence of comments are high level and generic, asking if my grandparents have passed, and telling me they are here and they send their love. This doesn't impress me at all, as most people my age would no longer have any living grandparents. But then she asks me about a young friend she says is here, a woman with a shaved undercut, long brown hair, and piercings. She goes on to describe, in detail, the appearance, posture, characteristic mannerisms, and personality of my friend Dagny, who died many years ago.

When I was twenty-five, I moved to Chicago and met her in recovery meetings. She was eighteen and struggling with drug addiction and alcoholism. She had already been to treatment more than once, as well as in and out of the hospital. She was the only child of a college professor father who was an intellectual and an atheist, and a hippy mother who was artistic and mystical. She was smart and full of passion. She made poems and paintings and music, and she loved to have long, deep conversations about life, death, the universe, sex, and everything in between. We were fast friends and I loved her like a little sister, and she had a crush on me.

"She's showing me a photograph of herself in the woods, sitting on a short wall maybe from a ruin, with her elbows on her knees and her palms facing up. She is looking into her hands, as though searching for a sign there in her palms."

This photograph exists. It is a paper photo from before the era of iPhones and I've preserved it in one of my journals. The shot was taken on a trip back to my hometown, which she came along for, and it was my twenty-sixth birthday. I was single at the time, and so I brought home friends instead of a girlfriend, and those friends were Dagny and Brita. The three of us were all newly sober and always together. Dagny was the youngest, me in the middle, and Brita was already in her early thirties. Brita and Dagny had gone for a walk in the woods and done a photo shoot with a film camera, as Dagny was experimenting with photography at the time, and taking a dark room class, which I took along with her. This particular photo where Dagny was seated and staring into her hands, she had given it a title. The title was intriguing and profound, though she offered no explanation for it, which was characteristic of her personality.

The title was: "Give Yourself to Your Hands."

If I had to guess based on the psychic medium's appearance, I would say she is a corporate type. Her long brown hair is impeccably styled, and she is dressed in a sleeveless blouse revealing strong shoulders. Her collarbones come together gracefully and shadows pour into their valleys like liquid. When she speaks, her hands move, and the shadows drain out, fill up, drain out, fill up. Her teeth are straight and white and the whole package screams corporate executive.

After describing Dagny, she goes on to describe my paternal grandfather. She sees him in his wire-rimmed glasses and sweater vest, smiling as he shows her a big stack of books. She asks me if he smoked a pipe. I'm not sure, but a pipe fits my recollection of him — and he did have a large library, which he left to me. After the session, I text my dad. My father replies: *Yes, he smoked a pipe most of his life, though he quit when I was a teenager.*

11.

Maria returns from her walk, and I tell her he's close to the end. She immediately weeps and rushes to his side. "I love you so much, Ewald, do you hear me? I love you so much." The nurse arrives to wash him and asks if I can stay and help her turn him. She grabs a leg and I lift an arm and it's like lifting a wet boulder. The foul smell of trapped sweat wafts up from the yellowed sheets beneath him and I'm careful not to breathe through my nose. Ewald groans and the nurse works the dirty sheets into a ridge beneath him, sliding another clean sheet in its place, also bunched into a ridge. We move to the other side and lift him again. The nurse yanks out the soiled sheet and I help her pull the clean one into place beneath him. We struggle similarly through changing his t-shirt and all the while, the nurse chatters on about the process of death, how he's showing signs with his sweat and clammy skin, his irregular breathing. "This is a dying man," she keeps saying, and I wonder if that's a useful thing to say. Maria continues to profess her love and spill her tears on his face, and he groans in response. The nurse suggests we give him morphine, and Maria does. Shortly after, he sleeps deeply, and the nurse is gone. The room is silent but for his growling snore and the soft sounds grief makes.

12.

The water falls from the wet beard of my dog to the wooden floor and this time, I let it sit. It's only water, after all, clean and translucent molecules of hydrogen and oxygen bound in an interlocking pattern making a liquid eroding a solid and evaporating into gas, lighter than air. A life in linear time is an erosion, and evaporation. An overlapping of hospice clients, of memories, ideas, boundaries in relationships, friends or lovers, sounds or silence, an overlapping of liquid and solid. It's not lining up. A life lived in linear time is not linear: it's circular, and my dog's beard water on the floor is proof. Further proof, her round bright eyes fixed on me, inviting forgiveness.

There is a light ache in my muscles from having exhausted them by my activities of the day, which recalls all the previous times I've exhausted them, which foretells all the future times I will do likewise. The rhythm of my pulse draws out the now, makes it have beats of succession, makes it contain just then, and about to come. It is a dynamic, sparking, poised blank. Poised to be filled, to be given to creation as whatever she wants it to be. My next move could be anything: jump up and scream, run out and kiss the first woman I see, roll onto the floor and moan, or simply still myself and listen. To the sound of a plane flying overhead, a familiar sound grounds me in my coordinates, the location of a body in latitude and longitude, in space and time. The chirping birds, the train approaching or departing. It's the sound of the train that holds the most reassurance. Its high, distant blaring is the call of my deep down now, the one poised, the blank that sparks. An empty space that sits on the summit of history, not just since I've been born but since generations before that and even farther back, all the way to the primordial ooze, and farther still, to the progenitor of everything that would ever be packed into in a dense, hot, vibrating, exploding ball. All of that underneath this opening, this now, poised for what's to come, when I depart the now into the next moment, the next now, the next opening to fill. Perhaps it is the coming and the going of the train, its simultaneity echoing the buried call, the low, knowing voice that can't be heard except as it reverberates in the slow blare and rolling boom of the train.

Birdsong has a similar effect, yet wholly distinct. One bird's song, the mourning dove, calls to memory my childhood, where the sound of crickets at night was the counterbalance, now missing. What that mourning dove does for me is nests that childhood nostalgically into my time torn free, my right here, right now. Departing, even as it arrives.

The world is full of companions.

13.

It is ten minutes away from Friday the 13th of May and the forty-fifth anniversary of my birth. I was born on a Friday the 13th and this seems significant. Not because I am in San Miguel de Allende again; not because my baby boy soulmate Dimas is spinning his small body on the floor imitating a breakdance with a little crowd circled around him, I among them, and my close friends among them; and not because we dance together, close together, whirling and everything whirling around us, lamps furniture walls floor, like a disc on a spindle, like the records spun by the DJ, a guy mixing beats with two turntables just like in a movie, and it feels as though we are in a movie, and I'm high on my own bright fantasy; and none of this is why Friday the 13th feels significant.

Rather, it feels significant because the ancient tribe that settled near here thousands of years ago in a high desert next to a fertile canyon built a pyramid deliberately designed to plant the moon two times each year right into its center on the horizon. They had a lunar calendar with thirteen seasons of the moon. What this makes me think of is time past, and I imagine the twirling room as a going back, back in time to a point at which I can do it differently. But there is nothing to redo, because the circling, spinning trajectory of life has landed me here, breathless and opened by joy, possessed by this one round moment with my hands in the air and my eyes on my hands and the giving of myself over to them once more.

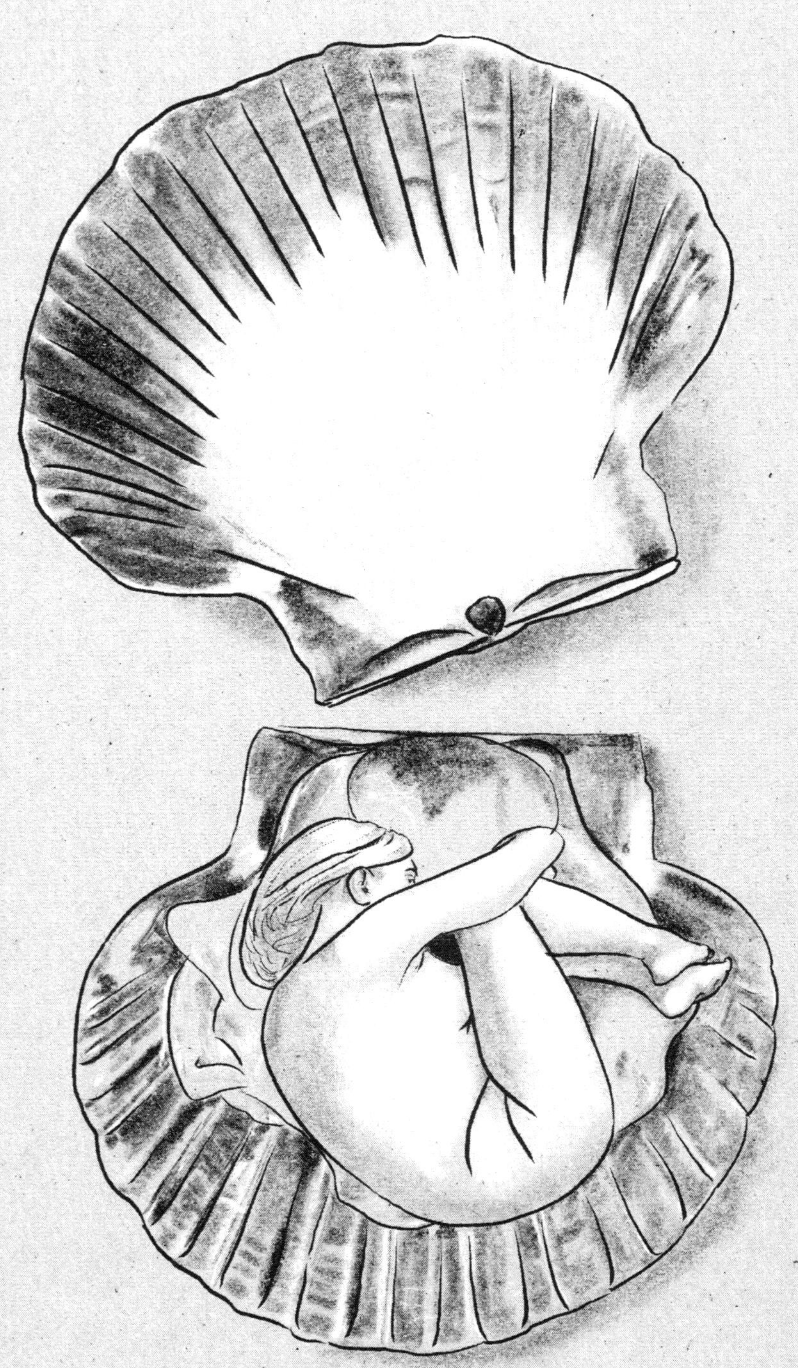

High Biscuits

"How are you today, Josie?" I am walking into the dark room where she sits, swallowed by a gray upholstered armchair with absorbent pads beneath her. She is dressed in a nightgown. I take in her swollen legs and feet against the gray carpet. Her pale face with wisps of curled white hair around it. Her dim eyes.

"I'm feeling a little skittley."

"Do you want some Reiki?" I ask, sitting on the floor near her feet like a child gathering close for story time. I place my hands on her legs just above her knees, and her eyes focus on the same tattoo as always, the one on the underside of my left forearm — my self-love arm. She points at the design and asks, "Do you like it that way?"

"Yes," I answer. And I do, very much. There are stars and a moon to remind me that a night sky, and all the possibility and unfathomableness it represents, is always near. There are Toni Morrison's perfect words that, when I first read them, touched a deep pocket of truth in my soul: *You're your best thing.* And there is that shape. That shape I'm always findings in bits of earth. The universal symbol for love. Of course, Josie is drawn to this arm. And of course, I like it this way.

"Did you get any tattoos in your life?"

"Oh no, not if I can help it. I have enough problems with what I look like." She laughs and I laugh, but also feel sadness curled up round in my ribcage because even now, here at the very end, having come full circle, Josie believes she has problems with what she looks like. When we exit these bodies, these shells, this planet — is it mostly our programming that remains? All the misogyny soft-coded into our DNA continuing to operate, even as all other functions begin to cease?

These vulnerable shells, these bodies, they are not bivalves. They have no spiney hinged shells to protect the soft, wet insides. Here at the end, Josie's body seems soft as though she is all flesh and no bone. Nothing hard is left.

She taps her leg and chest and says, "Yeah, they're still here. They're moveable." I close my eyes and feel the heat in my hands, rising.

///

Forces, like dimensions, are prolific. There are more than we know about. They are nameless. I will name them, and maybe you will remember their names when you leave here and go about your life. For the third time in my life, I am single, and for the first time in my life, I want to be single. Perhaps I want it because I know it as temporary. It's a state, not a part of my being, and I will be part of a couple again, and it will probably be soon. Aware of this, I value this state. I value the forces attending this largely hidden dimension in my life, the one in which I'm single, and my most important relationship is with you, my reader. So far, my *you* has been a lover. A romantic partner. Now, as a single person, my *you* is you, and you are my one. My only. We are practicing telepathy, which is a force.

Each force is an exotic animal I catch a glimpse of: a flash of yellow, a sleek pink body, shiny orange feathered heads. They can never be captured in words, but they can be approximated in a clear enough way as to render something adjacent, something close, something that points to what they really are.

Hannah Gadsby wrote a monologue about how men named all the things. But not all. Men don't have the vision of women. As Bataille revealed, they feel dead when they come. When women come, we feel numbers, taste colors, see the stench rising off you like infrared light. Like steam. We hear it drumming the air like millions of tiny wings, lifting, lifting, gone.

///

The next time I come to visit, I'm returning from a trip to Philadelphia, which is Josie's hometown. I tell her this. Her eyes light up at the mention of the city where she was born. I wait, excited for what she'll say.

"How did you get up and down and all around?"

"On an airplane."

"Ampony? What's an ampony?"

"An AIR-PLANE."

"Oh. Do you make those and go in those? How does that work?"

I don't bother with an answer. I just settle myself in front of her on the floor like always and take one swollen foot into my lap and wrap it in my warm hands.

"I feel that," she says.

"It's Reiki," I say.

"Where we goin'?"

"We're staying here. I'm here to visit you."

Her eyes are filled with fear, her energy anxious. She wants to go and is afraid to go all at the same time. She's already mostly gone though. She's going, leaving this zapping, zipping, nervous shell behind.

///

When two objects are entangled in a certain way, classical mechanics is overtaken by quantum mechanics. It stops being a matter of what we can see and touch and feel against our skin and what we can smell and taste. It starts being a matter of particles inside atoms, inside light. Billions on the tip of a needle. It makes you question everything, including your identity.

Who are you?

Take a moment to answer that question.

Did you tell me your name? Your gender? Your race? What you look like? Your location in time and space?

If you told me this last, you would be close to the truth, because no two objects can have the same four coordinates. Nor can they have the same four quantum numbers, which are principal, orbital, magnetic, and spin.

The spin, is it positive or negative? Or both? Is it a superposition of both at the same time, like a coin spinning on a table with both its faces showing simultaneously?

Superposition states are reduced to a single state. State one and state two are entangled, their spins become opposed, one is up and two is down. Separation doesn't break the entanglement. And you can know the position of one by observing that of the other. Einstein called it "spooky action at a distance."

///

Now that I've arrived, Josie's son, Steve, gets up to leave.

"Steve-Dave-Mike! Where you goin'?" she calls, tiny torches of fear burning in her eyes. She calls the names of all three sons together, not knowing which he is, but knowing he is her son.

"I'm Steve, Ma! I'm gonna go out and do some things. Elizabeth is gonna stay here with ya," he shouts.

"But I want you to stay around me!" she calls, and tears spill from her eyes, the torchlight still there, wide beneath the wet. Like a toddler. Exactly like how my son used to tear up and say "*Back!*" as soon as I was about to leave because he wanted my temporary absence to be over already and for the part where I'm back to be now.

He wasn't there yet, with the comings and goings of this world.

And Josie, she's done with the comings and goings of this world.

I'm not sure how, but what this tells me plainly is not only have I not conveniently and easily healed; I'm not even in touch with who I want to be anymore. Belief in some impossible true love of the romantic variety stubbornly persists in my mind, my body, sloshing like soup in my pelvic bowl. I'm not woke enough, not whole enough, not healed enough, not free enough. My eyes have never been open enough. I can't make them see what I have never seen, leave alone what's impossible for them to see. This shows me that old, original wounding still throbs redly in my psyche, and I've slid back even farther than I was before my first love. And the metaphysical cliff of illusion of strength and wholeness I'm gripping — it's crumbling. Pebbles and chunks of dirt pellet my face.

It's you who whispers, right into my ear, how it happened that I was led so astray. It's the shine in Josie's eyes, you say, that reminds you. All those loves, all those other you's: "She" was a shiny object. All of them were. And all my wounded selves were entranced so the whole school of them swam toward her. And it's true, there is proof. The shine got caught in my hair and flaked off into my underwear. I find bits of it on the bathroom floor like glitter.

///

Josie wasn't always like a toddler. Far from it, she has a PhD in logistical engineering. She worked for the government with top secret clearance. She made whole books of detailed, illustrated instructions on exactly how to build nuclear submarines from scratch. Every bolt accounted for. The tensile strength of the metal. No detail overlooked. She had been the president of the society of logistical engineers.

I look at her through a haze of astonishment and I almost see the strong and brilliant woman she once was. "Josie," I say, "is that true? Did you do all that?"

She nods and points to the nurse. "She walks with her hands," she says, and nods knowingly.

"It's a nice day out today," the nurse says. "Maybe Elizabeth can take you outside in your backyard? Wouldn't that be nice? I think it's a high of low seventies."

"At least it's not high biscuits," Josie says.

We all laugh. I think of that children's book both my kids loved, called *Cloudy With a Chance of Meatballs*. I imagine the weather report in that book being, "Clear today, sunny with high biscuits around noon."

I convey this to Josie. I wonder if there's a quantum world somewhere with high biscuits.

"I have a scream inside me somewhere," she says.

///

A quantum synced clock in Cleveland, Ohio would be the exact same pendulum swing as one in San Diego, California. By looking at one, you can know the position of the other instantly. That information traveling instantaneously and faster than light, is a quantum entanglement.

The partial knowledge of the state of two distinct systems. When the observed state of one improves our knowledge of the state of the other, the two systems are entangled. But this isn't right. This is a mere approximation of what quantum entanglements actually are. There are intricacies of entanglement that can only be accessed via advanced mathematics.

It's not as simple as quantum states of spin up or spin down. One of four possible combinations between two objects. Add all states together to get the quantum soup of all states happening at once: a superposition.

An entangled system is one where an interaction with one part affects the whole. Doing something to one seems to effortlessly and instantly affect the state of the other, no matter how far away it is. These explanations still aren't quite right, mind you. They are all efforts to use language to talk about things that can't quite be described.

///

"I'm feeling kind of dribbling."

Josie flips through the pages feverishly, calls out the page numbers, 16, 17, 18, then closes the book and puts it on her walker, nodding. She's agitated today. Her son Steve and his wife have gone, replaced by another of her sons, David, and his wife, Evelyn. David and Evelyn are going to move her in with them. They've been clearing out her house. Slowly, all around her, objects are disappearing. Furniture, framed photographs, books. They tell her she is leaving, that her house she has lived in for decades, where she raised her three children and lived with their father, is going to be sold. She forgets, but maybe she also doesn't forget. Maybe the vanishing of the familiar is affecting her, turning up the volume of the fear and anxiety she now vibrates with. I give her Reiki, but she won't accept it. She swats my hands away. I stand and back up. She picks up objects within reach and throws them at me, a tissue box, a travel mug. I retreat farther and hide in the kitchen.

"Get outta here, you asshole! Leave!" She pushes her frail body up from the armchair and stands, wobbly on her legs. "Give me my walker," she demands.

"Josie, please sit down, you'll fall," I warn.

"No, I gotta go. I'm going!"

I walk toward her and she throws punches, her face is contorted with rage. I retreat again. She makes her way away from her armchair by holding onto an end table, then a chair, then a table. She inches her way toward me. I take my phone from my pocket to call David and tell them to come back, but there's no signal. I walk to the door to go outside and find a signal. David and Evelyn are still there, sitting in their car, about to leave. I call out to them.

"Can you come back, please? She's not OK."

They both burst from the car and run back into the house, finding her in the kitchen now, red-faced and panting, drooling. "Get away from me!" she screams, as Evelyn approaches her.

"Take her hand and help her," Evelyn shouts at David. He walks to her and she grips his arm, clings to him.

"Steve-Dave-Mike! Help me," she says like a child. "Help me, help me," she keeps saying.

"I got you, Mom, come on. Let's go back to your chair." He helps her walk back to her chair and she falls back into it with a crash.

"Don't leave me, take me with you, help me," she says. David pulls a chair close to her and takes her hand. "I'm right here," he says. She stares at him, wide-eyed, her pale face pearled with sweat. Rather than flushed as before, her face seems drained of color now, completely white, as though she is looking at a ghost.

///

Quantum mechanics demands a departure from classical thinking. It demands new tools for understanding or describing language and sense perception. But even with the latter, we can't understand what is observed without using language to explain. Entangled electrons act like little magnets, the two spins must be opposite. Like a pair of gloves. You can tell the handedness of the other glove by seeing the handedness of the first. The crucial difference is the gloves are always the same.

Entangled particles act differently when observed. By nature of the emptied house, you know about the existence of Josie, the body who inhabited the house for so many decades. By nature of Josie and her fearful eyes, you can tell there is a place she imprinted upon very strongly and has left. Most of Josie's possessions have now left her house, just as most of Josie has now left her body. But it is still warm and firing off confused signals, like misfirings, willy-nilly. Which means some part of her essence remains, as though in transition from one house to the next, one body to the next, one synapse to the next, one neuron to the next. Like a chemical messenger. Something with indivisible meaning. She's in transition. Part here, part somewhere else.

Similarly, if one observes you sitting there reading this book, one can deduce that there exists a writer who wrote the book now in your hands. Or, by observing me in the act of writing this book, one could deduce that there will exist a reader holding it in the future. We are entangled.

///

We are all one second and one inch away from losing our minds, losing our abilities, losing control and even the illusion of control. All this loss will happen in due time. It is promised.

Do you want to know the secret?

Come closer, and I'll tell you.

This vulnerability that lays itself bare at the beginnings and the endings of these finite bodies? We are always (and in all ways) just as vulnerable. All the way through the middle. These hard shells fashioned from thoughts and pride and misguided beliefs during the prime years of our bodies — they aren't real. These binaries we have invented — the toothy, spiny bivalve housings, the sparkling creatures of the depths — they are but illusions. They are as vapor.

The last time I see Josie, she is calm. In fact, she's asleep more than half the time I'm there, seated in front of her giving her Reiki. I suspect, although she is already mostly gone, her vacated body is a long way still from turning off. She had been so strong when she was agitated and violent. She looked nothing like the frail, dying person she looks like now. But then again, some people are strong right up to the last breath. Like June was, and June's mother, who, she told me, "took a step and died."

Perhaps Josie was calm because Steve was back, and perhaps his energy is calm and hers is merely a reflection or extension of his at this point. And when David had been here, he was not calm, and so she was also not. Or perhaps Steve gives her the medications she is prescribed, whereas David did not. Either way, I am grateful to have her peaceful here at our final meeting.

I am thinking of the last book I wrote, and how my editor had suggested I had too many characters who were friends cluttering up the story, many only appearing one time. She suggested I collapse them into one. The suggestion alarmed me. I didn't realize that was something I could do in a story based on truth, a memoir. But then, of course I can do that. I am the writer of the story. I have the power to collapse innumerable friends together into one, epic, decades-long, ride-or-die friend. And so I did. And the effect was somehow healing, as though that one friend was real. She was one beautiful friendship, with me through it all, drifting in and out, coming and going, ever present as the energy of *friend*.

It occurs to me this is what Josie's remains are doing with her sons. She has collapsed the three together into one energy of *son*. It occurs to me that I can do the same with the series of lovers I've had for the past twenty-nine years. I can collapse them into one long lifetime of a love. We have had all the relationships inside the one, broken and healed each other's hearts hundreds of times over, and emerged still together. Still as one.

This one epic love, I see now, is you, reader. You are the one who has been here with me, from the beginning until now, the end. Not that this is an ending. You will go on after this story, and I will too, and we will sleep and wake and walk and dream and love again. And we will die and be born and live and die again. And only occasionally will we glimpse it, in a moment much like this — the whole. The round whole and how it's collapsed. It's one circle and all the circles, all together, all at once.

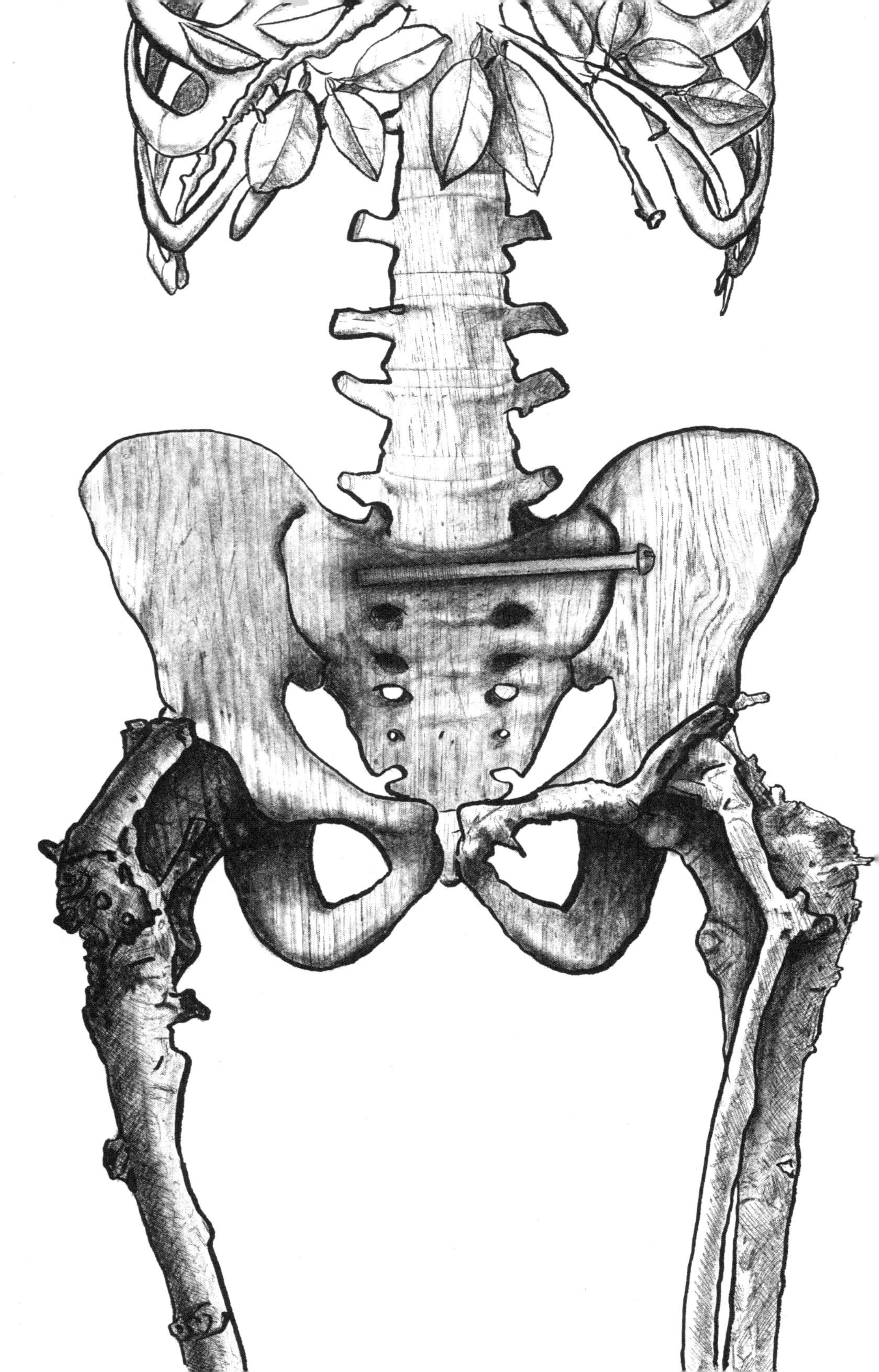

SCREWED

THIRTEENTH LOVE

You started like a constellation in the night sky: distant and mysterious and dreamy. Also, open to interpretation. On our second date, I gave you an early draft of this manuscript. In part to seduce, in part to weed out, in part to confess. In giving it to you, I felt the smallest touch of the kind of fear I will feel when it's complete and whole and out in the world, considered and handled by strangers, who for about twenty dollars will be able to take it home and read it and possibly discard it or dislike it or laugh at it. To make one's private musings public by choice — what a self-centered-and generous and crazy thing to do!

The experience of having this book out in the world may feel similar to having this large titanium screw removed from the base of my spine, connecting my sacrum and ilium. The removal will require surgery. Similarly, publication will hurt and be uncomfortable, and yet, my answer is yes. It is yes. Because, like the screw, it is this axis that is the access point for deep, paradoxical truths coiled up in the mitochondria of the human condition.

ax·is /ˈaksəs/

1. an imaginary line about which a body rotates.
2. a fixed reference line for the measurement of coordinates.

Because I say yes, someone will come along and say yes, and then another person will say yes, and that is all it takes. It takes one, maximum two people to say yes to this collection of words, thoughts, ideas, memories — to this story — for it to be published and disseminated into anonymous hands.

When that happens, will I feel naked? Will it feel the same as my clothes having been torn off by alcohol, and my hands having been held down by feet, and the weight of an older boy on top of me, one and then another? Will I feel screwed in the worst sense of the word?

Or, rather, will it feel like how it felt when I was six years old and in the first grade? I wet my pants one day in the hallway outside the classroom. We were lined up for recess or lunch or to go somewhere. With my soaked underwear and pants, I ran to the teacher and pleaded for help. What I didn't expect was that the teacher was mad. She asked in an irritated voice whether I wasn't old enough to control myself and get to a bathroom in time. I was shamed speechless. I lowered my head and looked at the floor and tried not to cry. Then it got worse. She roughly took me by the hand and pulled me into the nearby classroom, which was the second-grade class. There, in front of all the older kids in that class, she announced, "Might we borrow your bathroom? She had an accident and needs to get cleaned up." I remember a ripple of laughter across the room as I was escorted in toward the bathroom. The second-grade teacher looked as though she felt sorry for me and opened the door to the bathroom for me. I went in, and that's where my memory of the incident ends. Except that I wanted to never leave that bathroom. And a part of me never did. The part of me that feels ashamed of mistakes I've made — the messiest and most vulnerable humanness of me — and terrified of people knowing and seeing me this way. Somehow, to rescue my six-year-old self from that bathroom and give her a voice and bring her with me moving forward, I must say yes. Without shame, yes.

You said yes. You said yes to me. To us.

With you, I've been screwed in the best sense of the word, like never before.

Even before I met you in person, I saw the erotic energy sizzling and crackling off your skin like steam, one of the many manifestations of your impossible talent, your blinding shine. Up until that point in time, the one where I met you, there was a lot of breaking.

How did everything break so massively? Hearts and bones and vaginas. Breakups and crashes and birth. Human error. Technical error. The failure of judgment. Perhaps, if I hadn't taken things so personally, or if *She* hadn't been so avoidant, or both. If that driver had paid more attention, noticed my one headlight at dusk, and hadn't accelerated into my path when it was far too late for me to stop or swerve or otherwise avoid him. Maybe if I hadn't pushed so hard during childbirth, or if I had been given an episiotomy.

And yet.

Love returns and returns.

Bones heal.

Vaginas bounce back.

Most importantly: It all led me straight (or maybe curved) here, where I sit in a cab and the driver asks me if I'm coming home or visiting and I answer, "Coming home."

And though I *am* back home in San Diego, my answer feels false. Because home is your body. Because I have seen myself and this heartbreakingly beautiful world through your matchless eyes. And the effect is deep like bruises but with light. Like pain but pleasure.

How did this happen so fast?

I met you a mere three months ago. You, my thirteenth love. Three months pass quickly when you're falling in love. (Three months is an eternity when you're in pain, healing from all the breaking.)

My screw forms an important, intense connection in my body — one that has re-established my ability to stand, walk, and run. This connection has since healed, rendering the painful presence of the screw unnecessary. But the intense connection remains.

INTENSE CONNECTION

Once, when I was small and my mom was bathing me and the time came for her to drain the bath, I wept. Standing to climb out from the tub, dripping with bath water and tears, she asked me why I was crying.

"I don't know!" I moaned, and I didn't. She opened her arms then and I collapsed into them like a slippery pale seal returning to sea.

"You're just tired," she said, and held me. Wrapped me in a towel and squeezed me tight for what felt to my small body like a beautiful eternity.

Maybe this memorable encounter with my mom was what started my hunger for a different kind of intense connection.

I volunteered to help people in the process of dying partly out of this deep-seated longing for genuine and strong human connection. Since that first with my mom, there are several moments that stand out in my memory as among the strongest, most poignant.

MY OLDEST:

The third time I saw her, she was standing with her birth mom (my partner at the time) when I pulled into the driveway on my motorcycle. I watched as her two-year-old body jumped up and down, her small hands covering her face, a scream barely audible over the rumble of my engine. I turned off the bike and dismounted, took off my helmet, and as soon as I put it down, she ran to me and leapt into my arms shouting my name. Holding her, the spontaneous and overwhelming love I felt made me stagger a few steps backward. It was in that moment, I knew — we would belong to each other forever.

MY YOUNGEST:

As I gathered what negligible strength I had left for one more heave, it occurred to me that I would possibly die, that this last one would bring another whole body out through my pelvis and tear me completely in half and kill me. I cried, mumbling "I can't, I'll die," but it was as automatic and inevitable as the next inhale, and so I pushed and tore and broke and out he came, silent and grey and motionless into the room. There was a still moment, only held breath and heartbeats, before all the bodies in the room exploded into action. There was no letting the cord stay attached awhile; no immediate delivery into my arms. Urgent attempts to save my baby's life trivialized all these plans I had, as well as the long anticipation of learning the sex, which someone revealed by calling out the obvious, "He's not breathing!"

They whisked him away and I was furious with terror, screaming and crying, "Give me my baby! Give him to me!" I saw people gathered around him several feet away, suctioning out his airway with a small vacuum hose, and I had no coherent thoughts, only panic, and I could not speak, only shriek. The nurse stood between my baby and me, her arms stretched out to the sides as if trying to touch us both at the same time. She was reassuring me, "You'll hear him soon, listen, wait for it, he's going to be ok ..."

When his cry came, his body was placed on mine. On contact, he peed, the hot liquid spilling over my sides and mixing with the blood and fluids under me. The relief I felt was just as hot, just as liquid, and it washed me clean.

THE MAN WHO SAVED MY LIFE:

His face was mostly peaceful. The off-duty physician, who came to my rescue moments after I landed on the hood of the car and began to bleed out. He lifted the visor on my helmet, his face was close to my face, he shouted over my screams that he was a doctor, he calmed me down to slow my blood loss, he cut off my boot laces and tied a tourniquet. He promised me I would live. The calm on his face was well rehearsed, I could tell — a doctor's practiced expression for excellent bedside manner. Still, in my heightened state, I saw that expression crack. I saw it tinged with desperate fear that no matter what he said or did, I was about to die right in front of him.

THE CHAPLAIN IN THE TRAUMA WARD:

During the first several days I spent in the trauma ward at Scripps Mercy Hospital following my near-fatal accident, I was heavily drugged and terribly haunted by the gruesome image of my right hand torn off and hanging by skin, splintered blood-covered bone jutting out. This image visited panic upon me, panic that held within it a sudden and terrible knowing that I would soon die, and that this inevitable event had come prematurely and without warning. Nightmares featuring severed limbs and invading predators darkened my every dream-like hallucination. Much like the hospice volunteer I was later to become, a volunteer hospital chaplain visited me during that terror. She sat beside me, and I told her the story. It was the first time I told the story from the beginning to a stranger who had no context, only my badly bruised and broken and pieced together body in a hospital bed. When I got to the part where my right hand was hanging off, and when I formed that image with words for us both to vividly see, I shook and wept. It took a moment for me to realize that she, too, shook and wept. She lay a hand gently on my arm and simply cried with me.

YOU, MY THIRTEENTH LOVE:

It occurs to me that all my other strongest human connection moments were outside of the container of romance. And although I have had many strong attractions, and many intense sexual encounters, there is something very distinct and sure about you. There was our first date, when we sat in the grass in a park and fuzzy blooms from a flowering summer tree filled the air, floating like fairies or thick snowflakes all around us, but without the cold. The energy between us vibrated and sparked, premonitions of the storm to come, which would be filled with the lash of thunder and lightning and so much wetness.

With you, I have been screwed. Screwed like ecstatic orgasmic reciprocal dense pleasure so all consuming, it fills every individual cell with light to bursting. In particular, there was the night in Oakland when I strapped the cock you call *Big Guy* around my kintsugi bowl pelvis and thrust it deep into you. It was the conductor of our two big energies, the lightning rod between us. Our energies merged and swelled, vibrated, and sparked — it felt as though we were coming apart cell by cell and combining into one enormous helix of swarming bits of light.

You: vertical and perpendicular and erect — the y-axis to my x-axis. You fit perfectly against my body with your/my *Big Guy* inside you, in a way as to provide the perfect friction to make my wet clit sing. I watched as pleasure broke apart your face the same way pain must, the same way it doubtlessly broke apart mine. The sounds I made crescendoed in harmony with your perfect sounds until, at the very same moment, each of us died just a little, and came hard enough to finally, if temporarily, sate the hungry animal of our mutual desire.

The feel of you is as seismic as the earth cracking open. And in I fall, swallowed whole. Gone.

On top of you, I feel you healing my masculine, draining all the toxicity right out of him, and replacing it with powerful, creative, erotic light. And sealing the cracks with gold.

VARIATIONS ON SCREWED

In Which the Screw Is My Best Friend

I first saw it on the X-ray screen four years ago, and it alarmed me. "Why is it so big?" I asked the surgeon. "It's all the way across my back to the other joint on the opposite side!"

"That screw is your best friend," he said.

My sacrum and ilium had hit the pavement with such force that they came apart. This collision of my body with the ground is referred to as "Impact 1" on the police report diagram. "Impact 2" is the collision of motorcycle and automobile. "Impact 1" also resulted in my left thigh being torn to the bone.

The sacroiliac joints link your pelvis and lower spine. They're made up of the sacrum (the bony structure above your tailbone and below your lower vertebrae) and the ilium (the top section of your pelvis). There are bilateral sacroiliac joints in your lower back.

The word *sacroiliac* reminds me of celebrity couples who are renamed as combinations of their original names. Bennifer and Brangelina. When my sacrum and ilium "broke up," they were rejoined by this screw, which reestablished and preserved their crucial connection and mutual dependency. The surgeon was right — this screw is definitely my best friend.

Four years after seeing the screw on the X-ray screen, it is time to have it pulled from my pelvic bones. I've always been drawn to paradox, and I relish the idea that I want the screw out of my body and much as I fear living without it. How can you live without your best friend? I want it gone because, for more than four years now, there at the base of my spine, my best friend has stabbed me in the left ass cheek. My best friend also seems to rust in place whenever I sit for too long, hurting like hell the moment I stir. I can almost hear it squeaking and creaking, as though it needs oil. Movement makes my best friend stop hurting, but even when it doesn't hurt, it always makes its presence known. We have that in common. Despite these pains and this haunting constant presence, my screw is my best friend for restoring the integrity of my pelvis. My screw is my best friend for allowing me to walk, to run, to swim, to hold weight, to screw. My screw is my best friend for making me whole again.

In Which the Screw Is the Spine of a Book

In the chiropractor's office, I see a model of a pelvis and the bottom half of a spine. It shows the simple architecture of the bowl: That structure in my body has been so damaged. So rebuilt. So healed. I can see and touch the places that were injured. On the model, it's clear to me how it would have opened, unhinged like a book. How, under the right pressure, it is anatomically possible for these bones to open that way, like a book. How, when it opened, it revealed a story long concealed. A story as old as my farthest, earliest ancestor. A story I have carried in these very bones from the moment they were formed inside my mother's body.

This hinge, filled with screw, will soon be empty. Replacing its weight will be light and buoyancy. The fibrous and malleable strength of muscle and marrow.

In Which the Screw Is an Antenna

Ancient Mesoamerican tribes considered the sacrum to be both a bodily and cultural keystone, connecting it with body symbolism, mystical experience, and cosmology. It represented an access point for the translocation of shamans, spirits, and deities between worlds or levels of the cosmos, and they believed it played a significant role in reproduction, fertility, and reincarnation. Similarly, the word *sacrum* comes from the Latin *os sacrum*, meaning "holy bone," which in turn is translated from the ancient Greek, so called either because it was the part of the animal they offered in sacrifice or because it was considered the seat of the soul. This screw, my screw, was driven right through my sacrum. For four years it has been my own personal pelvic keystone. It tunnels through the bone right where the marrow should be, and I believe it acts as an antenna, receiving signals just the way a radio antenna intercepts sound waves. I believe this because since its acquisition, I have felt more sensitive to the energy of other people, and also to the energy of places.

In Which the Screw Is Ability

I dreamed I had ridden to Costco on a motorcycle and gone inside on crutches. I was post-surgical and not allowed to put weight on my left leg at all. I was supposed to allow it to heal. But inside Costco, I found myself without crutches, teetering on one leg. Hopping slowly forward on one foot, it was too hard, especially because my arms were filled with bulk portions of food. I walked, placing all the forbidden weight on the now screwless joint connecting my sacrum to my ilium to my hip. It collapsed easily, and my body crumbled, and I used my arms to army-crawl myself out of the rubble of spilled apples, frozen burritos, and pretzels filled with peanut butter.

Cut to a scene in which I was laying in the back of a cargo truck. My children were waiting with me while my co-parent, *Papa*, shopped inside. I was relieved that he was doing what needed to be done for our family while I was incapacitated. I was surrounded by the most vulnerable members of my family: my children. Inexplicably, my mother was there. Maybe simply because my body is of my mother as much as it is of the earth. Maybe because in real life, she, too, is vulnerable in her older age and failing body.

In my waking life, I am about to have surgery to remove the screw. For a time, I will not be allowed to put any weight on my right leg. I worry about my children, whether they will be re-traumatized by witnessing their faster-than-the-speed-of-light mother tripped up, in pain again. I worry it will shuttle them back to the time in their life that they came so close to losing their mother at such a young age.

ARE YOU A STRANGER WHO HAD ME?

When my son was four, he asked very deep questions. After a veritable monologue in which I explained that he should not talk to strangers and not go anywhere with strangers — and in which I explained how to get help if strangers were asking him to escort them somewhere — he remained silent for a time. We were walking down the sidewalk, hand-in-hand. He looked up at me and asked, "Are you a stranger who had me?"

His insight charmed and thrilled me. When he was even younger, I chastised him for breaking a rule, and he replied by saying "I'm sorry." I said, "It's OK," and he replied, "There, I fixed your rule." His response had charmed and thrilled me then, as well.

At the base of this question, this logic, is the question of what makes a person a *stranger*. If you assume that the defining characteristic of a *stranger* is that you don't know them, then everyone you know was once a *stranger* before you came to know them. So I answered yes. Yes, I am a stranger who had him.

When I was a pubescent child, I loved gaining the power to make people cry. The power to cause a strong emotional response. The power to rise above the station of *stranger.* To affect a person in service to evidence that they cared, or that I'd tamed them.

In the classic story of *The Little Prince,* a man encounters a boy in the Saharan desert after crashing his propeller plane there. After trying and failing to draw a hat for the boy, and then a sheep, the boy tells the man the tale of how he came to be here, on this Earth, and he wonders why there are no people. In the end, the boy has tamed the man, and to be tamed is to be attached. To be special to someone is to be tamed by them. What matters is not the body among thousands of bodies, the flower among thousands of flowers, the tree among thousands of trees — these are but shells. In the end, the man learns that what matters is invisible.

THIRTEENTH LOVE

Are we screwed? I wonder if we'll ever find a way back from this place. This impossible place we find ourselves, delivered here by what seems to be nothing less than the bone, flesh, and blood hands of destiny. This place where the love I've lost is the price for this love I found. A price I would be happy to pay again and again and again. Impossible because it defies time and space. Impossible because, how could we ever go back and settle for anything less than this?

I don't ever want to come back from this place. From this love. If it must end, I would like to fall off the edge of the earth. Fall into the sky, fall all the way to the raging sun. Add my light to its infinite light. Add my heat to its deadly heat.

If the sacrum is sacred, if it supplies access and serves as an axis, if it is a portal for other worlds, then the hole made in my sacrum by the absence of the screw might remind me of the holes we make. The angles of our geometry and the curves of our bodies blast holes through these three axes, these three dimensions, showing us glimmering, electric, shimmering new ones.

I want you too much. So much, it feels as unsafe as my tiny breakable body gaining velocity through space, flying precariously among metal and concrete. I want you too much, and there is no smaller amount of wanting I could anymore bear. Any less would feel futile.

And even as I hold these paradoxical truths in both my lungs like held breath, I can't help but think of that time in Mexico, when I lay naked on top of you and sang along with Bruce Springsteen these words: *I wanna read your mind and know just what I've got in this new thing I found.* This, after, inexplicably, a hummingbird beat its small body against the windowpanes of our bedroom. It could not have been real, except we both witnessed it. We stared, held in rapture and disbelief, at what was there right in front of our eyes. It had to be a sign. Was it wanting to nurse off the sweetness of us, like sugar water from a hanging feeder? Wanting our ambrosia? Our nectar?

You immediately looked it up on the internet: *What is the meaning of a hummingbird?*

You, standing there, still clothed at the time, read aloud that a hummingbird symbolizes joy and pleasure and love. Me, already naked in the bed, impatiently waiting for you — I said, "Then come over here and give me joy and pleasure and love."

The time between then and now feels like a blink.

Passing time with you feels so easy and too fast.

When you are away, my longing for you aches in all my joints, those with metal and those without. These bones and this flesh are as embedded with my longing for you as the bones previously broken are embedded with surgical-grade steel.

You are an artist, and you make illustrations on your tablet. Watching you, I learn that the software you use to draw has digital pens and brushes that are programmed to mimic the way ink or paint lays itself on a canvas. You have a digital brush entitled *Wet and Easy.* You jokingly but seriously told me how this reminds you of me. I laugh, like I often do in your presence, filled with joy as you make me, and explain to you that this reminder — it's true and is entirely your fault.

I want you by my side as I heal from the surgery. Because lying with you feels as fervent and peaceful as prayer.

TIME CAPSULE

When I was with my first love, we made time capsules for one another. We wrote each other a letter and gathered a few special items that were representative of the time and that would fit into the manilla envelope, then sealed the envelope and wrote on its face: DO NOT OPEN UNTIL YOUR FORTIETH BIRTHDAY.

She kept her time capsule for me, and I kept mine for her. At the time, we were about seventeen, and forty seemed as far in the future as the first signs of life on earth seemed in the past. I don't remember what objects I placed in the time capsule, nor what I wrote in the letter, and I lost the envelope somewhere along the way. Of course, I never received that time capsule from her, and I'm now forty-five. And yet. What matters is invisible.

The screw in my sacred sacrum along with the rest of the metal in my body is like an object placed in a time capsule. Since my unforgettable brush with death, I often imagine my future cremated corpse — the pile of ashes decorated with metal. My body will obey the alchemy of fire and be reduced to dust, but the hardware won't. And so I long to have that surgical grade steel, that for now reinforces my skeleton, placed into the hands of alchemists and smelted into decorations for the bodies of my descendants. Bracelets, perhaps. Or rings. Fine, thin, delicate rings as numerous as the cascade of lives I will leave behind when I die, the little bodies, the little shells.

And each body, each hand, each shell can have a ring. A ring from the lightning rod of the universe, the receiver of the signals from the cosmos — the screw. A ritual ring symbolizing the eternal round of birth, life, sex, death, rebirth. Little circles, all dense with meaning, all in a row.

Acknowledgements

This book has been a passage through intimate terrain where sex and death meet — through the places that most scared me, the moments that most harmed me, as well as those that inspired and grew me, expanded my range. I went directly to the beating heart of each body I wanted instead to shy away from because I've learned that these are the ones that hold the rawest, most powerful truths of our humanity. In braiding together personal confessions, scientific research, and philosophical insights, I have not only sought understanding, but also liberation. To tell these stories, to name these animals, is an act of feminist resistance — a refusal to be silenced or sanitized. In a world that too often shames the body, censors the erotic, and denies mortality, these pages are a reclamation.

We are living through a pivotal moment — one of collective reckoning and possibility. To share truthfully, vulnerably, and without apology is to push against the tides of repression and conformity. This book joins the chorus of revelatory voices insisting that our experiences matter, that our stories — especially those shaped by trauma, pleasure, grief, and joy — must be heard. Feminist consciousness grows not just through theory, but through lived truth. May these words inspire others to speak, to listen, and to rise — together.

Thank you, first and foremost, to Nicole Roberts, the extraordinarily talented artist and brilliant creative mind who made custom, revelatory, thrilling illustrations for each of these essays. When we combine our minds and talents, what gets produced is infinitely greater than the sum of those parts. I am honored and humbled to be part of this powerful collective we are building together.

To steal her words, "oceans of gratitude" to the vastly talented and smart Lidia Yuknavitch. Thank you for creating spaces for people like us to "make art in the face of fuck." Meeting you was alchemy and this book was shaped in part by your Body of the Book workshop and the nurturing I received from you and that group.

Thanks to Emily Rapp Black who gave the collection necessary editorial love and suggested a structure that became its skeleton, without which, it may never have filled with spirit as it did and animated, claiming this life all its own.

Thanks to Jennifer Racioppi for the year you spent with me and the time you invested into knowing and understanding me on the deepest level. The insights and direction and inspiration you offered helped me to not only bring this book out into the world, but to graduate to a new level of self care and love I desperately needed. Thank you for expertly midwifing that process for me.

Thanks to my early readers, Alan and Ellie and Leslie, and the valuable insights you offered. Thanks to my badass launch team at Jaded Ibis Press and beyond: Carmen, Dominique, Dana, Julia, Angela, and team. You have believed in this book and in me and I have needed this faith to lift me forward. Thanks to Jeanine Rose of JeanAudio for the hours spent listening to and recording this audiobook, and for the conversations we had between those sessions. I now believe it was your distinct and unique energy that was required to couple with mine in just this way and bring this format of this book into being.

Thanks, finally, to the beautiful souls who have touched mine to inspire these stories and contribute to the shape of what I am today, here, now. I will always love you.